HUMAN CHROMOSOMES

HUMAN CHROMOSOMES

An Illustrated Introduction to
Human Cytogenetics

AUDREY BISHOP and PATRICIA COOKE

The United Sheffield Hospitals
Centre for Human Genetics

WILLIAM HEINEMANN MEDICAL BOOKS LTD
LONDON

First published 1966

Filmset, Printed and Bound in Great Britain by
Bookprint Limited, London and Crawley

Foreword

Since the discovery in 1959 that there were individuals who did not possess the normal male (XY) or the normal female (XX) arrangement of sex chromosomes, great interest has been aroused in human chromosomes. So rapidly have advances been made and so voluminous has the literature become that it is difficult for those not continually in touch with cytogenetics to keep abreast of events. Furthermore, since so many recent publications are reports of complicated chromosome abnormalities or accounts of the minutiae of advanced laboratory techniques, it is even more difficult for ordinary interested doctors, students or nurses to assimilate the rudiments of the subject; they are apt to be left floundering in a maze of karyotypes, metaphases, translocations, satellites, centromeres and deletions without truly understanding the words used let alone their significance.

In our work in the paediatric and gynaecological disciplines of medicine we are aware of the confusion which exists in the minds of medical and nursing personnel on chromosomes. We often hear the view expressed that an elementary book does not exist which explains them simply. We are therefore particularly pleased that Mrs. Bishop and Mrs. Cooke, both experienced cytogeneticists, have remedied this deficiency. Their book provides basic information on the subject of human chromosomes; gives a simple description of normal chromosome behaviour; it demonstrates how the commoner abnormalities arise and what clinical features are associated with them. The high quality of the many illustrations ensures that everything is easy to follow. In writing this book the authors have provided a much needed elementary account of a complex subject which is assuming considerable importance in so many fields of medicine. We are delighted to write the foreword to such a necessary publication and we wish it all the success it deserves.

April 1966

C. J. Dewhurst, M.B., Ch.B., F.R.C.S., M.R.C.O.G.
R. R. Gordon, M.C., M.D., M.R.C.P., D.C.H.

Contents

Chapter		Page
1	The cell, and its activity	1
2	The culture of human cells for investigation of the chromosomes	8
3	The identification of human chromosomes	10
4	Changes in chromosome number	17
5	Structural changes	36
6	The X chromosome	48
	Glossary of clinical terms	55
	Index	56

Acknowledgments

We are deeply grateful to our colleagues for their helpful suggestions during the preparation of the manuscript.

A.M.B.
P.C.

All the preparations, photomicrographs and diagrams were made by the authors

Chapter 1

The cell and its activity

The cell is the basic unit from which are built the organs and tissues that form the human body. The appearance of cells in different parts of the body may vary but basically they are made up of two parts, nucleus and cytoplasm. The cytoplasm is a complex viscous fluid bounded on the outside by a thin selectively permeable membrane and on the inside by a similar membrane surrounding the nucleus. The nucleus is made up of a network of chromatin in a liquid medium called the nuclear sap. Nuclear chromatin is that material which, at some part of the life of the cell, may be stained with basic dyes such as lacto-aceto-orcein, Feulgen or haematoxylin. It is made up mainly of molecules of deoxyribonucleic acid (DNA) which carry chemically coded genetic information, responsible for the manufacture of proteins and the reproduction of the cell.

During its life the cell passes through several stages of activity. The stages are named for descriptive convenience but the transition between these stages is a gradual one. The so called resting nucleus is in the stage known as interphase. During interphase the cell is physiologically active. One form of physiological activity is synthesis of ribonucleic acid (RNA) and protein. Molecules are assembled along the DNA of the chromatin in a specific sequence determined by the genetic information carried by the DNA to form molecules of RNA. These molecules of RNA migrate to points in the cytoplasm at which protein synthesis takes place. Since this RNA is responsible for carrying genetic information from the DNA to sites of protein synthesis it is referred to as messenger RNA. Other molecules of RNA in the cytoplasm pick up amino acids and transfer them to the sites of protein synthesis. This is transfer RNA. These amino acids are assembled in a sequence determined by the messenger RNA to form specific proteins.

It can be seen that the type of protein produced is dependent on the genetic information carried by the DNA. The units of information carried by the DNA are known as genes. A single gene may be responsible for the production of a specific protein. Any change in structure of this gene may result in a change in the type of protein produced leading to an upset in cell metabolism. Similarly, any change in the frequency of this gene may alter the quantity of the protein produced, leading to imbalance in cell activity.

The cell also undergoes a phase of reproductive activity. This process may be considered in two parts, replication and division. At replication, another form of physiological activity which also takes place at interphase, the molecules of DNA in the chromatin double themselves. All through interphase the strands of chromatin are diffuse and only palely staining so that they cannot be seen with the light microscope. Just prior to cell division however, the chromatin threads, which have now replicated, condense to form discrete, densely-staining bodies known as chromosomes. In normal human somatic tissue there are 46 chromosomes. These are made up of 23 genetically identical pairs of chromosomes or homologues. Where both members of a homologous pair are present in a cell it is said to be diploid and the chromosome constitution is represented by the symbol 2n. Diploid cells may undergo two forms of division, mitosis or meiosis.

Mitosis is associated with the growth and maintenance of somatic tissue. It is a process whereby daughter cells are produced of exactly the same chromosome constitution as the parent cell. Meiosis is a form of division which only occurs in the gonads during the production of gametes (ova or sperm). It may also be called reduction division since it is a process whereby the number of chromosomes in the gamete is reduced by half. Gametes contain only one member of each homologous pair of chromosomes, 23 in all. This is the haploid number of chromosomes and is represented by the symbol n. On fertilization, two haploid gametes fuse to give a diploid zygote from which a new individual develops. The cells of the new individual contain one member of each homologous pair of chromosomes from each parent.

During division the cell is mechanically active. The onset of cell division both in mitosis and meiosis is characterized by the condensation of the diffuse chromatin network. When distinct threads of chromatin can be seen the cell enters prophase. From this point on the processes of mitosis and meiosis diverge. In mitosis (Figs. 1.1 and 1.2), as prophase proceeds, the chromosomes appear as double structures made up of two genetically identical chromatids joined at a

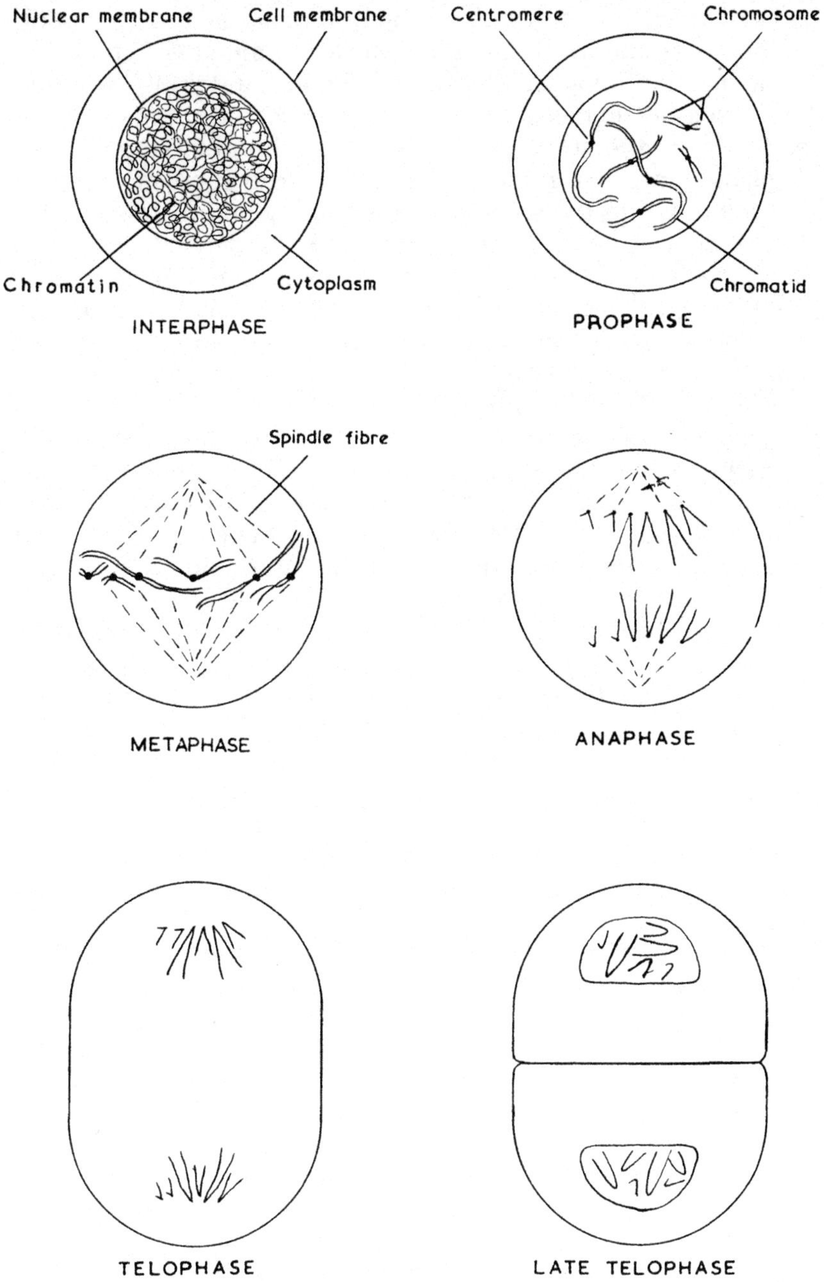

Fig. 1.1. Diagram of the stages of mitosis.
In this, and in all subsequent diagrams, only 3 pairs of chromosomes (n=3) are shown, representing the normal human complement of 23 pairs (n=23).

point called the centromere which appears as a constriction in the chromatin thread. Prophase terminates when the nuclear membrane breaks down. Fibres stretching between the poles of the cell then begin to coalesce from the nuclear sap to form the spindle. At metaphase the next stage in cell division, the centromeres of the chromosomes become associated with the spindle half way between the two poles. Since all the chromosomes are now condensed and have come to lie in the same plane, metaphase is an advantageous time for studying chromosomes. Detailed investigations of the chromosomes at this stage may reveal abnormalities of number and morphology which may be associated with imbalance of cell activity.

The centromere now splits but the two halves remain in association with the spindle fibres. These first contract, pulling the two halves of the chromosome apart and then the whole spindle elongates, thus further separating the two groups of chromatids. This stage is anaphase and the separation results in the formation of two groups of genetically identical chromatids. The groups of chromatids are then organized at telophase into nuclei while the cytoplasm divides to form two daughter cells. These daughter cells then return to their interphase state during the early stages of which the daughter chromosomes replicate once more.

Meiosis (Fig. 1.3) differs most profoundly from mitosis during prophase. Instead of being scattered at random through the nuclear sap, the pairs of chromosomes become longitudinally associated while still in a relatively uncondensed state.

The associated pair of chromosomes which is called a bivalent is made up of four chromatids. While the two chromosomes are associated in this way the phenomenon of crossing over may take place resulting in exchange of genetic material between homologues. The sites of crossing over are called chiasmata. Towards the end of prophase the four chromatids tend to fall apart, being held together only at the points where crossing over has occurred. While pairing and crossing over take place the chromosomes gradually contract until, at the end of prophase, they reach maximum condensation. At the end of prophase the nuclear membrane disappears and the bivalents come to lie mid way between the two poles with the two centromeres of each bivalent oriented along the long axis of the cell. At anaphase the two halves of each bivalent move to opposite poles thus ending any temporary

Fig. 1.2. The stages of mitosis as observed in human fibroblasts.

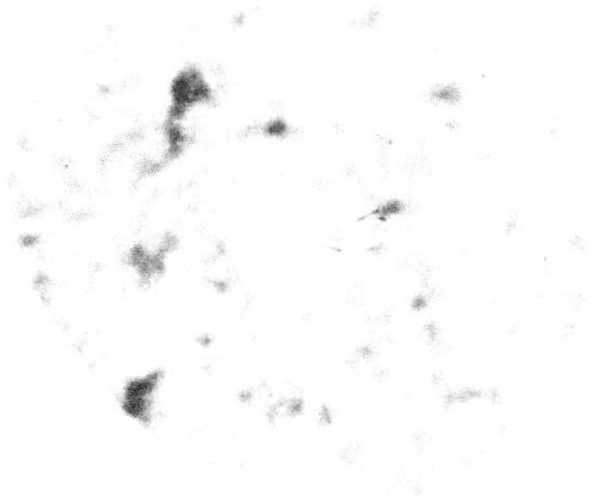

Fig. 1.2a. Interphase.

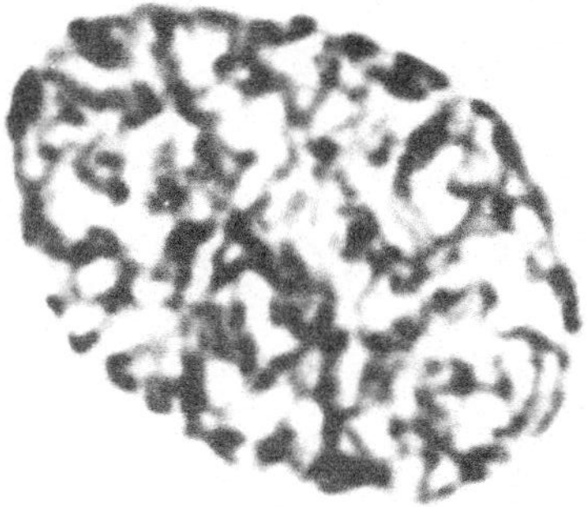

Fig. 1.2b. Early prophase.

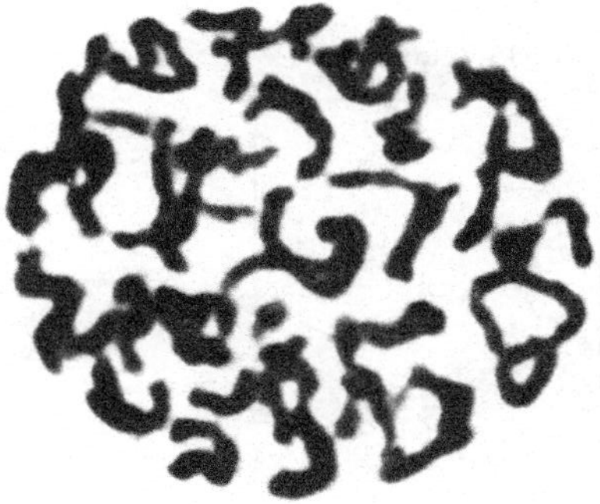

Fig. 1.2c. Late prophase.

Fig. 1.2d. Metaphase.

Fig. 1.2e. Anaphase.

Fig. 1.2f. Early telophase

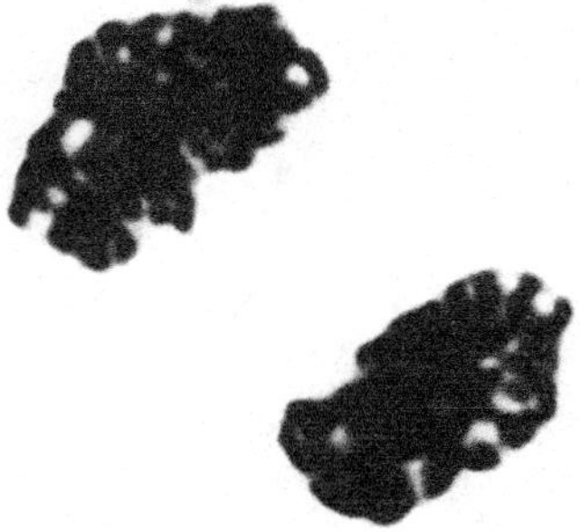

Fig. 1.2g. Late telophase.

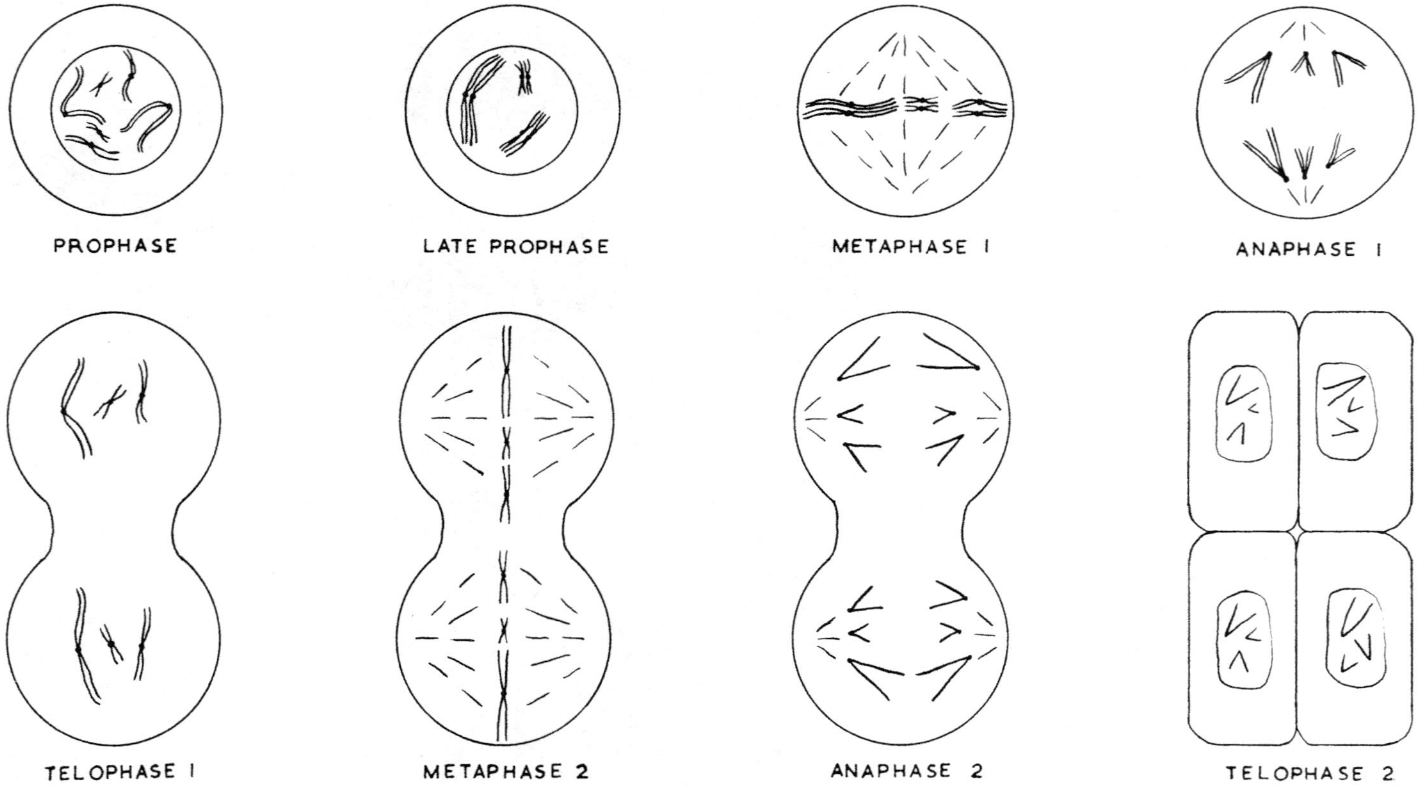

Fig. 1.3. Diagram of the stages of meiosis.

associations caused by the crossing over process. The anaphase separation results in the formation of two groups of chromosomes, each group being made up of half the original chromosome complement with one chromosome from each pair. A reduction division has been effected. At telophase these haploid groups of chromosomes may be organized into nuclei while the cytoplasm divides, but in some organisms this stage is omitted and the haploid group of chromosomes proceed directly to the next part of meiosis which is an equational division of mitotic type. Metaphase, anaphase and telophase follow as in mitosis, resulting in the production of four cells each containing the haploid number of chromosomes. To distinguish conveniently between metaphase, anaphase and telophase of the first reduction division and the same stages in the equational division the subscripts I and II are used. These two divisions together constitute the meiotic division.

Chapter 2

The culture of human cells for investigation of the chromosomes

Human chromosomes are most easily studied at the metaphase stage of mitosis. To obtain such cells, a sample of tissue is grown outside the body under sterile conditions, in a suitable growth medium at 37°C. When the cells are actively dividing, colcemid is added to block the cell cycle at metaphase. The colcemid inhibits the formation of the cell spindle and so prevents the chromosomes dividing on the metaphase plate and being drawn to the poles. These cells are then swollen by suspension in a hypotonic solution. This swelling separates the chromosomes from each other and aids their eventual identification. The cells are fixed in acetic alcohol before being made into slides for cytological examination. The chromosomes may be spread apart by dropping the fixed cell suspension onto a slide covered in cold water and the slide allowed to air dry. The preparation is stained in a suitable stain such as lacto-aceto-orcein and mounted for examination.

Many different tissues in the body including blood, bone marrow, skin, muscle and gonads may be used in this way for making chromosome preparations. Chromosomes in mitosis, the most generally studied division, are derived from somatic tissues. Chromosomes in meiosis, which are more rarely studied because of the difficulties of obtaining specimens and good preparations, are derived from gonadal tissue.

The techniques used may be illustrated by descriptions of blood culture, bone marrow preparations and skin culture. A sample of peripheral blood is mixed with heparin to prevent clotting. Phytohaemoglutinin (an extract of the bean) is added to aid the separation of the white cells from the red cells and to transform certain of the white cells, which do not normally divide in the blood, into cells which will divide outside the body. The white cells are usually separated from the red by centrifuging. They are mixed with a synthetic medium containing the necessary nutrients and antibiotics to prevent infection. The culture is incubated at 37°C. for

2–3 days when a wave of mitosis occurs. The cells are then harvested and made into slides.

Chromosome preparations are derived from bone marrow by a direct method which arrests the first cell division after removal from the body and can be used for any tissue, such as tumour, which is dividing rapidly in the body and can be easily dispersed. The specimen, is transferred into isotonic saline containing colcemid. After two hours incubation the cell suspension is harvested as before. These cells give a truer picture of the cell types present in the body than cells which have been grown for longer periods, as artefacts may arise in culture. This is particularly important in studying cancerous diseases such as leukaemias, where abnormal cells typical of the disease may be lost in culture, in favour of the normal cells.

Skin culture is of value where blood is difficult to obtain and may give further information about the chromosome complement in cases of mosaicism (p. 21). Skin culture is also of particular use in research where studies of chromosome behaviour are made easier by using cell lines which can be maintained and subcultured for long periods. A piece of skin about 1 mm. diameter is aseptically cut up and fastened inside a sterile glass vessel by clotted plasma. A compound medium, which is usually a mixture of synthetic medium, human serum and chick embryo extract is added and changed at regular intervals. After about two weeks there is a thick growth of fibroblast cells produced by the original specimen, which adhere to the glass surface. These cells may be removed by suspension in trypsin solution, centrifuged and then resuspended in new medium. When several subcultures have been made in this way, the cells may then be trypsinized and harvested.

Chapter 3

The Identification of human chromosomes

The chromosome complement cannot, with present methods of analysis, be arranged in 23 pairs but may be divided into at least 12 subdivisions. The chromosomes are arranged in descending order of length and, where the lengths are similar, they are further divided according to the position of the centromere. Those which have the centromere at approximately the middle of their length are termed metacentric, those with the centromere between the middle and the end are submetacentric, and those with the centromere nearly at

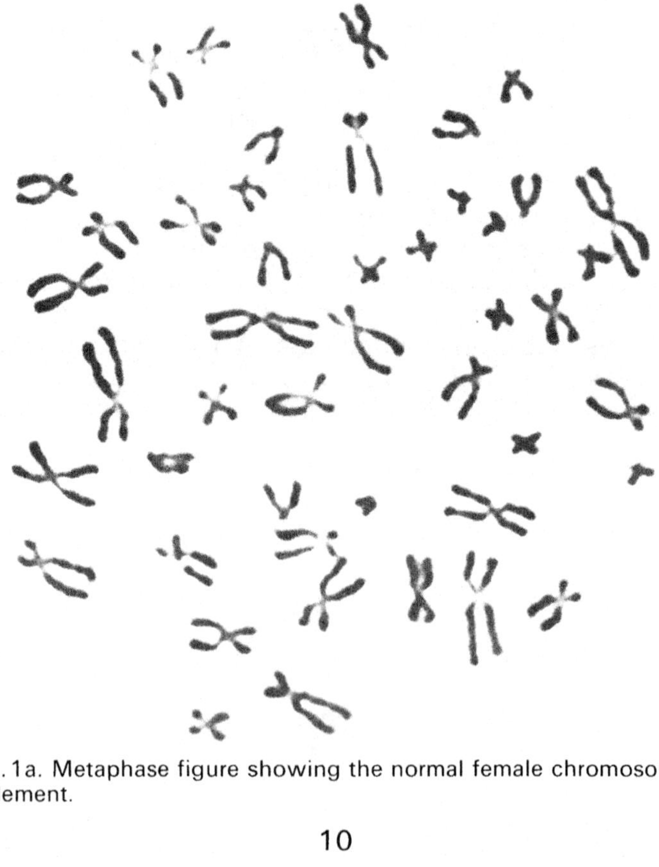

Fig. 3.1a. Metaphase figure showing the normal female chromosome complement.

the end are acrocentric. Having arranged the complement by these criteria, similarities between adjacent chromosomes allow them to be grouped into pairs or groups of similar morphology. Fig. 3.1a shows chromosomes from a normal female cell. Fig. 3.1b shows the chromosomes arranged in this way to form a karyotype. The sex chromosomes are called X and Y. The other 22 pairs are called autosomes. The normal female has two X chromosomes and no Y chromosome. The X chromosomes are difficult to distinguish from the 6–12 group by the standard technique.

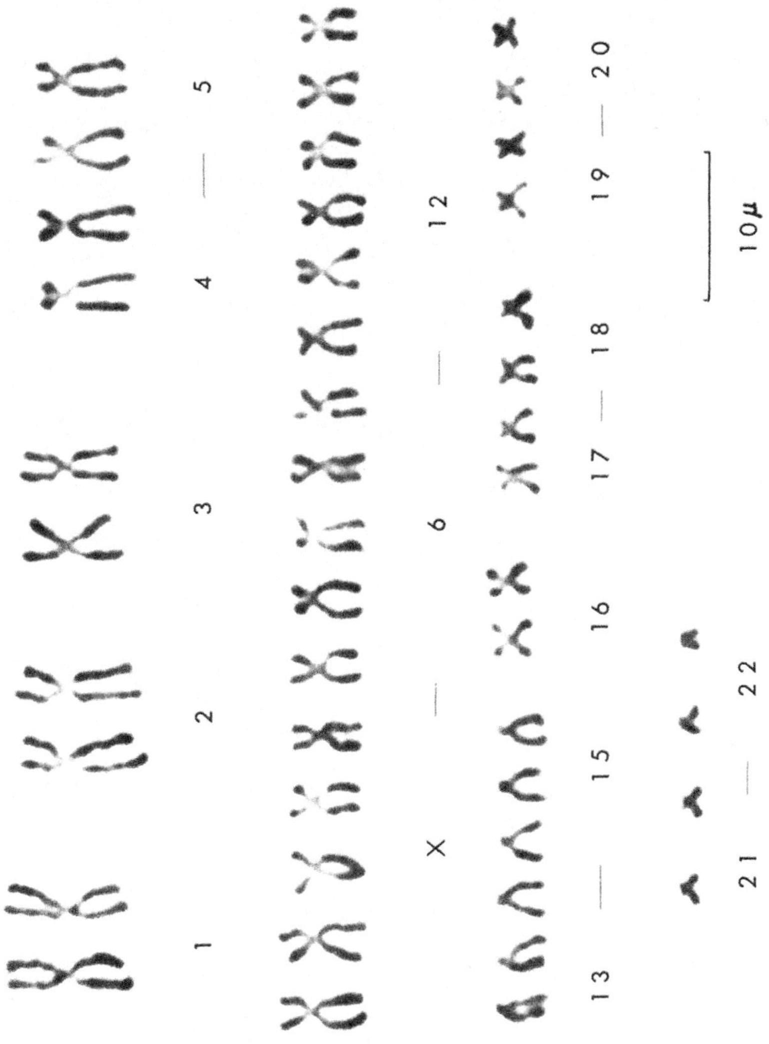

Fig. 3.1b. Karyotype of the normal female chromosome complement.

The normal male (Figs. 3.2a and 3.2b) has one X and one Y chromosome. The Y chromosome is usually distinguishable from pair No. 21 and pair No. 22 by its having more parallel chromatids, a more terminal centromere and by the absence of satellites (p. 14).

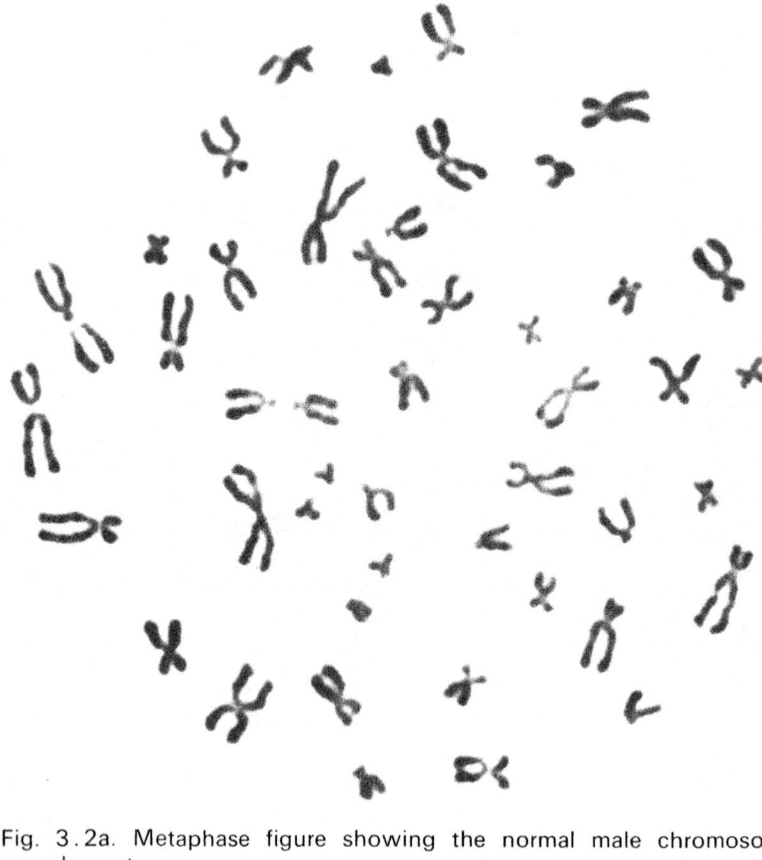

Fig. 3.2a. Metaphase figure showing the normal male chromosome complement.

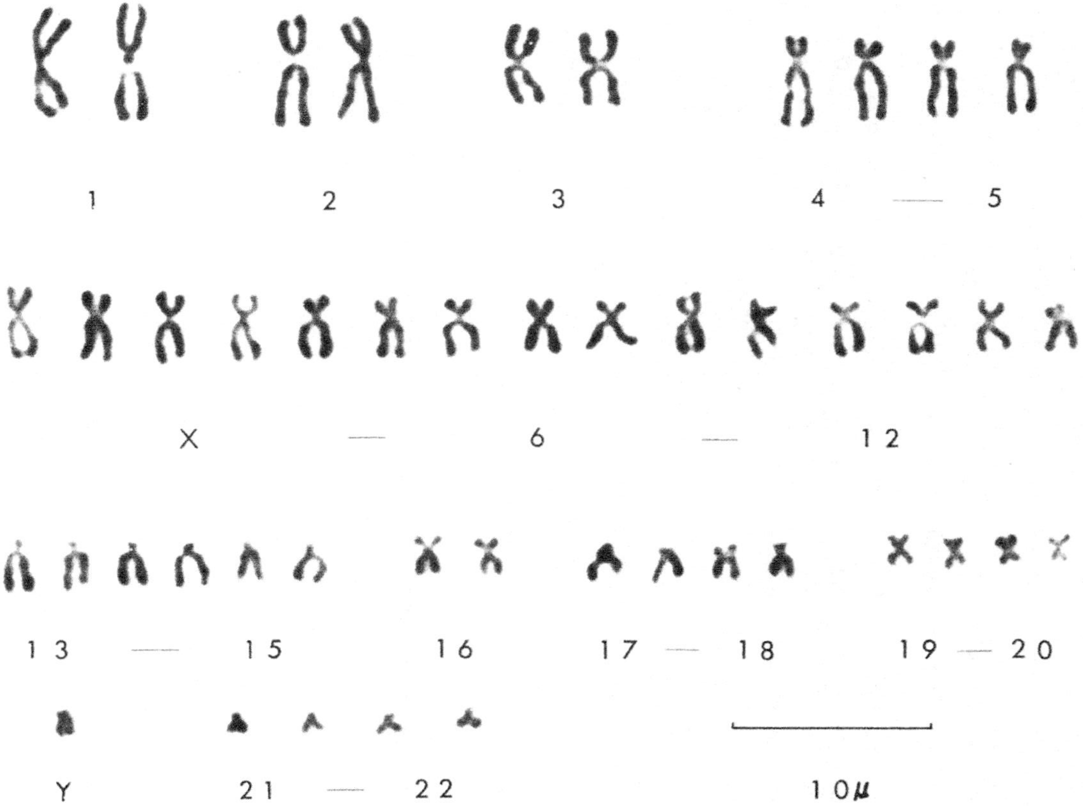

Fig. 3.2b. Karyotype of the normal male chromosome complement.

Secondary constrictions are narrow, stretched parts of the chromosomes which are more palely staining (Fig. 3.3). They are called secondary to distinguish them from the primary constriction of the centromere. They are apparent in specific regions in some of the chromosomes but only under certain conditions of preparation. They may be accentuated by special techniques such as growing the cells in a calcium-free medium. As the positions of the constrictions are specific, some chromosomes may be further identified by their presence.

Satellites are knobs of chromatin attached to the short arms of the acrocentric chromosomes by stem-like secondary constrictions (Fig. 3.4). The acrocentric chromosomes are often observed to be associated by their satellites. The presence of satellites and this association are aids to analysis, more clearly distinguishing the acrocentric chromosomes.

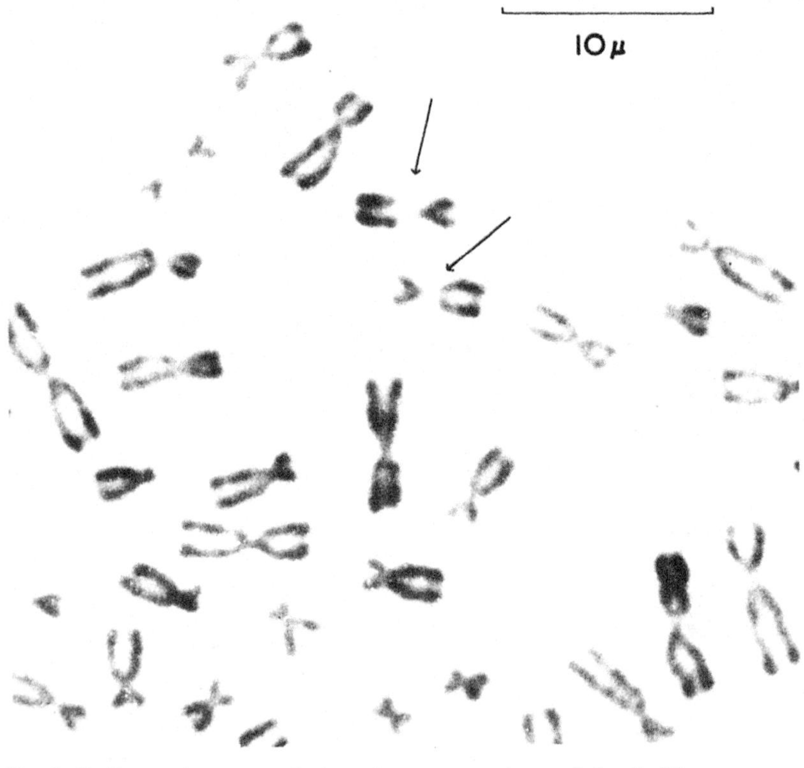

Fig. 3.3. Secondary constrictions in two members of the 6–12 group.

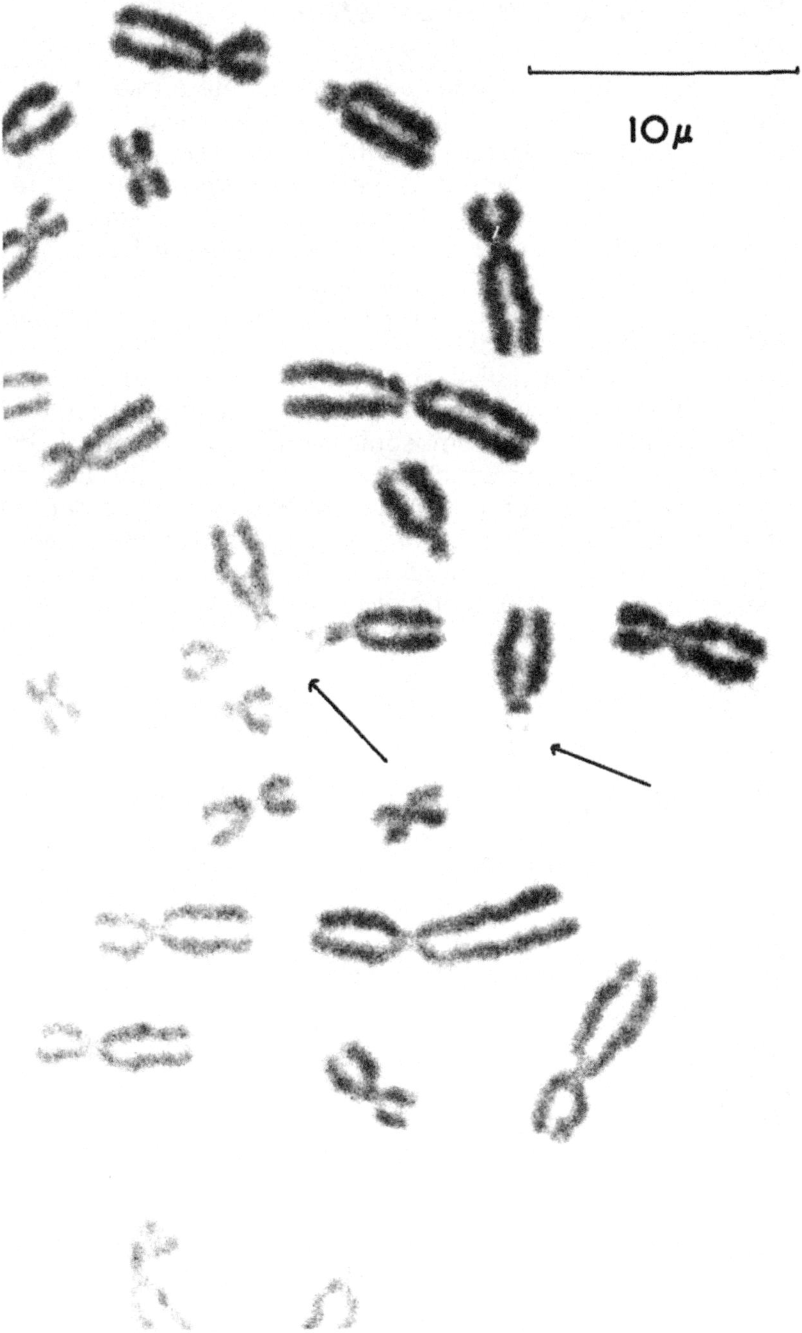

Fig. 3.4. Satellites on members of the 13–15 group and association between two of these chromosomes and a member of the 21–22 group.

The use of radioactive isotopes

When a chromosome replicates during interphase, it synthesizes new molecules of deoxyribonucleic acid (DNA). If growing cells are supplied with tritiated thymidine they incorporate this compound in their newly synthesized DNA. Tritium is a radioactive isotope of hydrogen emitting β particles, and the presence of the isotope may be located by placing special photographic film over the specimen. The film, still in contact with the slide, is developed and fixed. This autoradiograph then shows a pattern of radioactive labelling visible as silver grains over the chromosomes (Fig. 3.5). DNA is synthesized in different chromosomes and in various parts of a chromosome at slightly different times. Only those segments which were replicating at the time of treatment with the isotope will be labelled. Therefore a pattern of replication may be established which is used to characterize certain chromosomes, particularly those which cannot be distinguished by morphology alone.

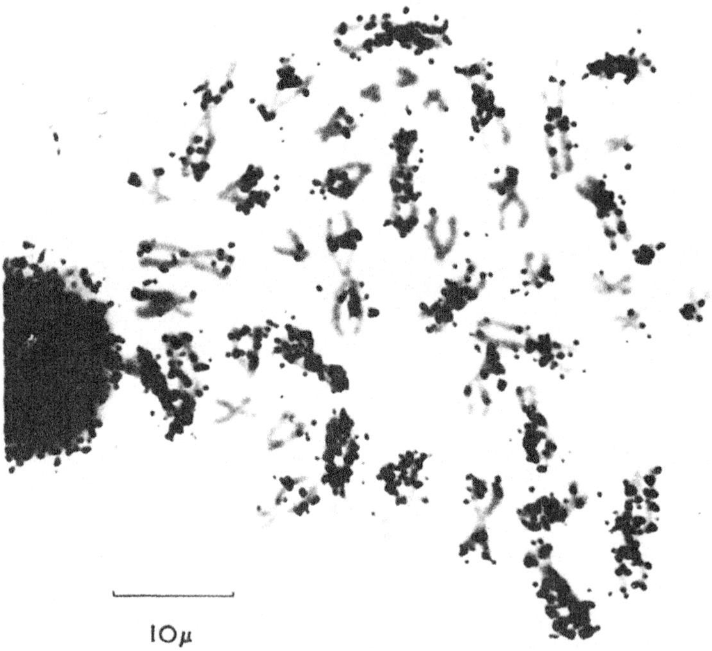

Fig. 3.5. Autoradiograph showing the pattern of radioactive labelling over a metaphase figure.

Chapter 4

Numerical changes in the karyotype

The processes of mitosis and meiosis are responsible for the accurate transmission of genetic information. To maintain the accuracy of this transmission the whole mechanism of cell division is under rigid control. However, errors can occur.

The two commonest upsets of division which result in changes in the number of chromosomes present in any cell are chromosome non-disjunction and anaphase lag. These are both anaphase phenomena and may take place either during mitosis or meiosis. Normally at anaphase of mitosis and anaphase II of meiosis, one chromatid from each chromosome moves to the opposite poles of the cell. At anaphase I of meiosis one chromosome from each bivalent moves to opposite poles of the cell. Non-disjunction is said to occur when this regular and equal separation does not take place, both chromatids of a chromosome or both chromosomes from a bivalent moving to the same pole. The chromatids or chromosomes fail to disjoin.

The results of non-disjunction are variable and dependent on the time at which this upset takes place.

1 Non-disjunction at Anaphase I of meiosis (gametogenesis) (Fig. 4.1).

A single non-disjunctive event occurring at this stage gives rise to 2 cells, one with 24 and one with 22 chromosomes. If the equational division of these cells is uneventful the four products of the whole meiotic cycle will give rise to gametes of these two cell types which, on fertilization by a normal gamete with 23 chromosomes, will produce zygotes with respectively 47 chromosomes (duplication of genetic material) or 45 chromosomes (deficiency of genetic material).

An individual with an extra member of a chromosome pair is said to be trisomic for this chromosome and an individual with only one member of a chromosome pair is said to be monosomic for this chromosome.

2 Non-disjunction at Anaphase II of meiosis (Fig. 4.2).

If a single non-disjunctive event takes place in one of the

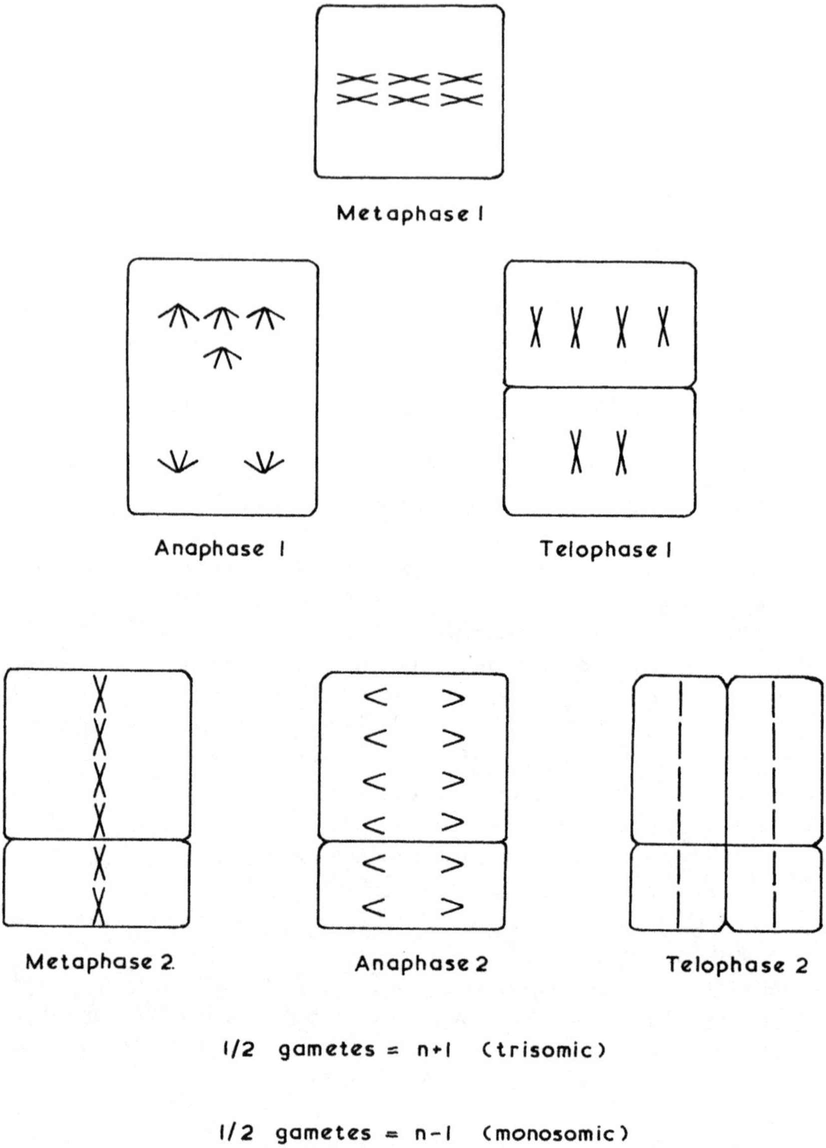

Fig. 4.1. Non-disjunction at Anaphase I of meiosis.

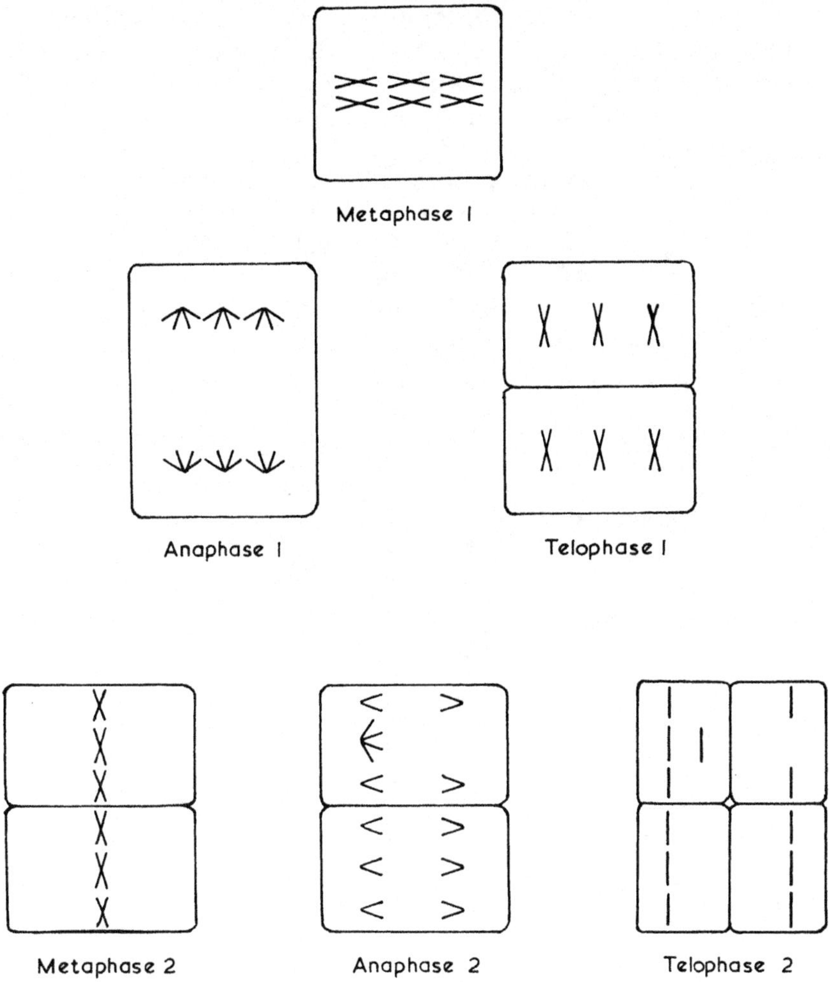

1/4 gametes = n+1 (trisomic)

1/4 gametes = n−1 (monosomic)

1/2 gametes = n (normal constitution)

Fig. 4.2. Non-disjunction at Anaphase II of meiosis.

products of the first reduction division of meiosis, then the four meiotic products may be of three cell types, i.e. with either 24, 22 or 23 chromosomes.

3 Non-disjunction at Anaphase I followed by non-disjunction at Anaphase II (Fig. 4.3).

The two abnormal products of non-disjunction at Anaphase I may further show non-disjunction at Anaphase II. If non-disjunction takes place in the cell with 24 chromosomes the products will have 25 and 23 chromosomes, or if it happens in the cell with 22 chromosomes the products will have 21 and 23 chromosomes.

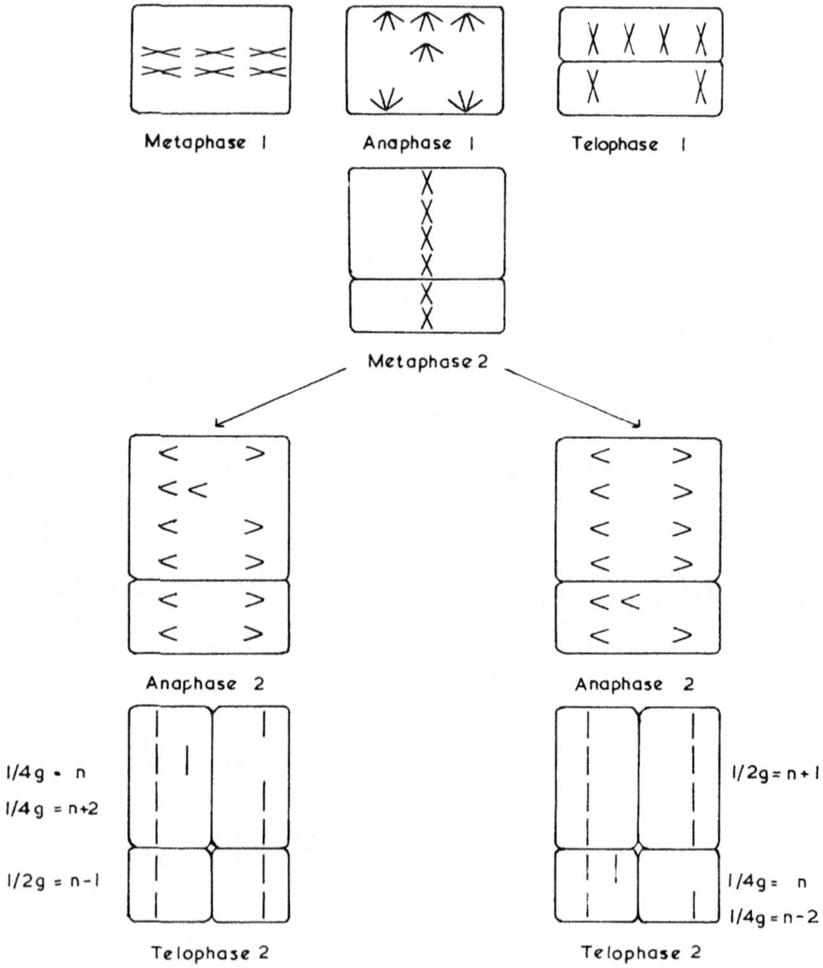

Fig. 4.3. Non-disjunction at Anaphase I followed by non-disjunction at Anaphase II of meiosis.

4 Non-disjunction at mitosis (Fig. 4.4).

Non-disjunction may occur at anaphase of mitosis during the development of the embryo and the effect of such non-disjunction critically depends on the time at which it takes place. If the upset takes place very early in embryonic development, then the whole of the embryo may be made up of abnormal cells. However, if it occurs later in development, then only some cells of some tissues may be affected.

Individuals in whom two or more cell lines with different chromosome complements are present are called mosaics.

The other common upset of division is anaphase lag (Figs. 4.5a and 4.5b). This is said to occur when a chromatid or chromosome fails to move to either pole at anaphase. It 'lags' behind, is not included in a nucleus and is lost. When this happens one of the products of division may be deficient for this lagging chromosome.

Daughter cells deficient in genetic material (showing monosomy) are apparently relatively inviable since they are rarely observed. There is one exception to this, the X chromosome, and there are special reasons why the X chromosome is unusual in this respect (*see* Chapter 6). The most common numerical upsets observed are, therefore, trisomies. Most mosaics, except those involving the X chromosome, are made up of a mixture of normal and trisomic cells.

In humans, individuals have been found with numerical abnormalities of the chromosomes which must have arisen from misdivisions during gametogenesis or embryonic growth and development. Karyotypes and descriptions of individuals with numerical changes are given in the following pages.

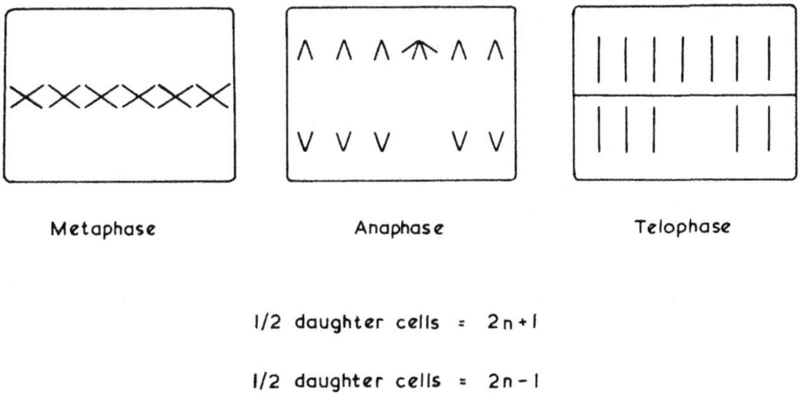

Fig. 4.4. Non-disjunction at mitosis.

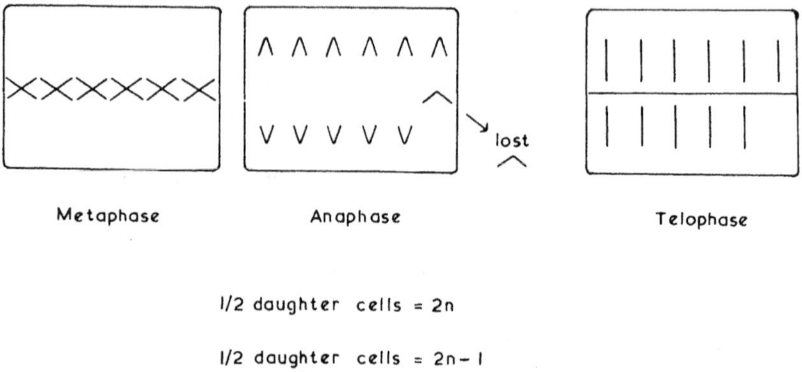

Fig. 4.5a. Anaphase lag at mitosis.

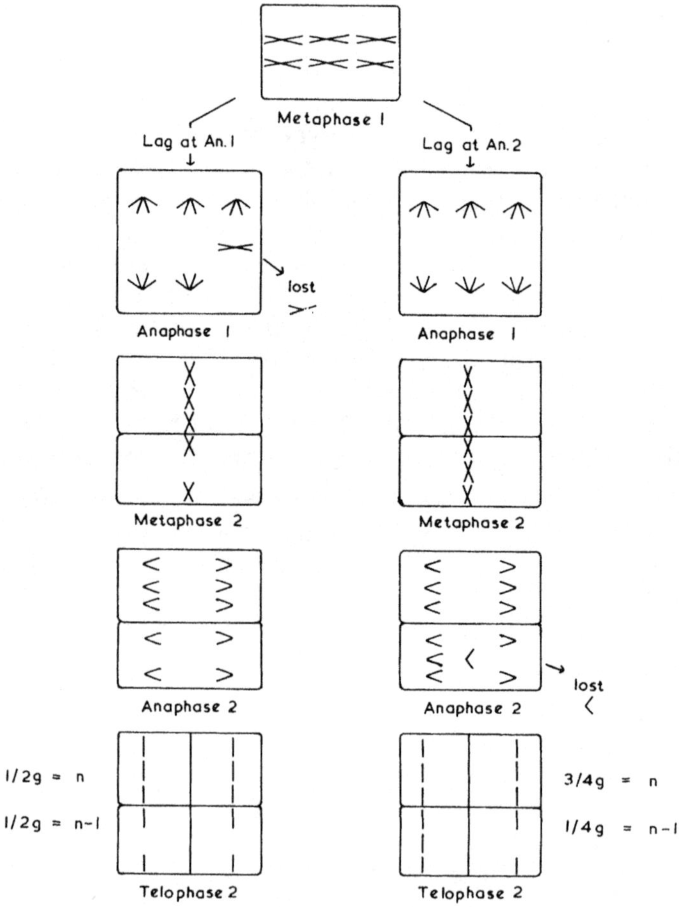

Fig. 4.5b. Anaphase lag at meiosis.

1 Mongolism (Fig. 4.6).

The clinical condition known as mongolism (Downs syndrome) is associated with mental subnormality. The physical features frequently but not invariably observed include a characteristic facies with epicanthic folds and oblique palpebral fissures, flattened occiput, Brushfield spots in the eyes, fissured tongue, short incurved fifth digits, abnormal dermatoglyphic patterns and, in the newborn, hypotonia.

This condition has been found to be associated with the presence in some or all of the cells of the individual of extra genetic material from the 21–22 group. This extra material is most often present in the form of an extra chromosome in the 21–22 group. The individual is conventionally considered to be trisomic for chromosome 21. This is the most common autosomal anomaly observed in man.

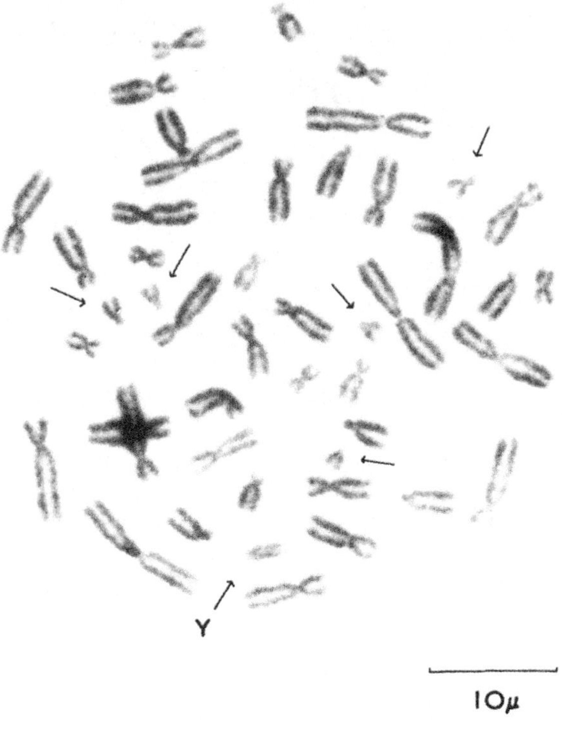

Fig. 4.6. Metaphase figure from a mongol with 47 chromosomes including 5 members of the 21–22 group. It is interesting to note that in this case there is an exceptionally long Y chromosome present.

2 Trisomy 17–18 (Fig. 4.7a and 4.7b).

Individuals with an extra member of the 17–18 group are grossly mentally retarded and have multiple physical malformations including micrognathia, low-set ears and characteristically clenched hands with overlapping fingers. Again the extra chromosome may occur in some or all of the cells of the affected individual.

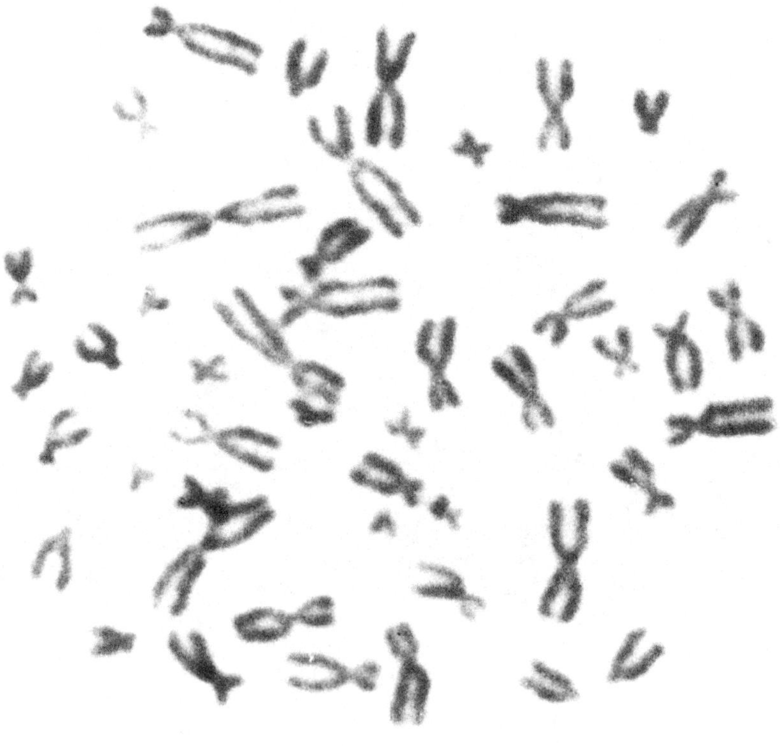

Fig. 4.7a. Metaphase figure showing 5 members of the 17–18 group (47 chromosomes).

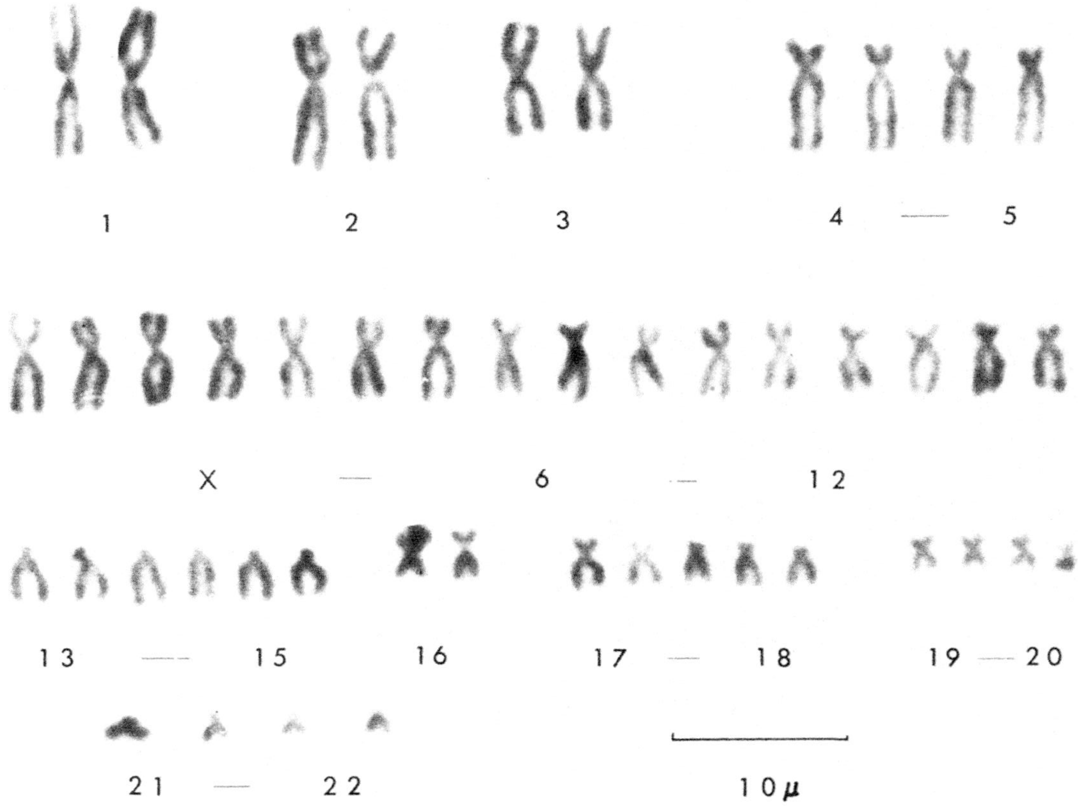

Fig. 4.7b. Karyotype from Fig. 4.7a.

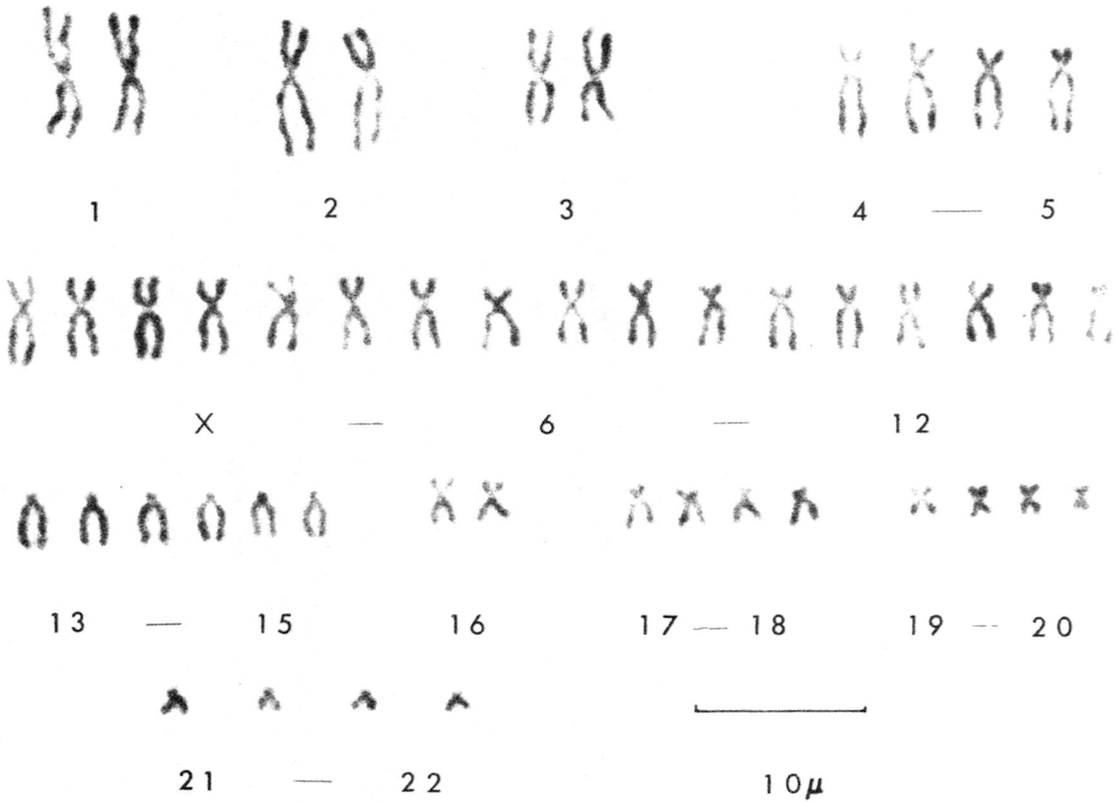

Fig. 4.8a. Karyotype from Fig. 4.8b.

3 Trisomy X (Figs. 4.8a and 4.8b).

The physical features associated with this anomaly include mental subnormality and sometimes poorly developed secondary sex characters with amenorrhoea. If the extra X chromosome occurs in only some of the cells it is likely that the reciprocal product of the non-disjunction, an XO cell line, will be observed in the resulting mosaic.

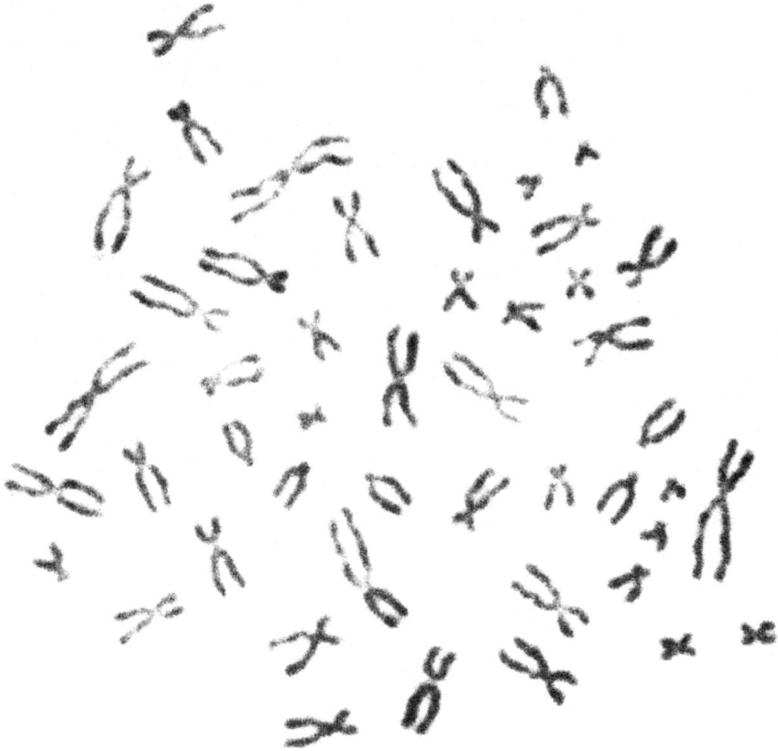

Fig. 4.8b. Metaphase figure showing trisomy X (47 chromosomes).

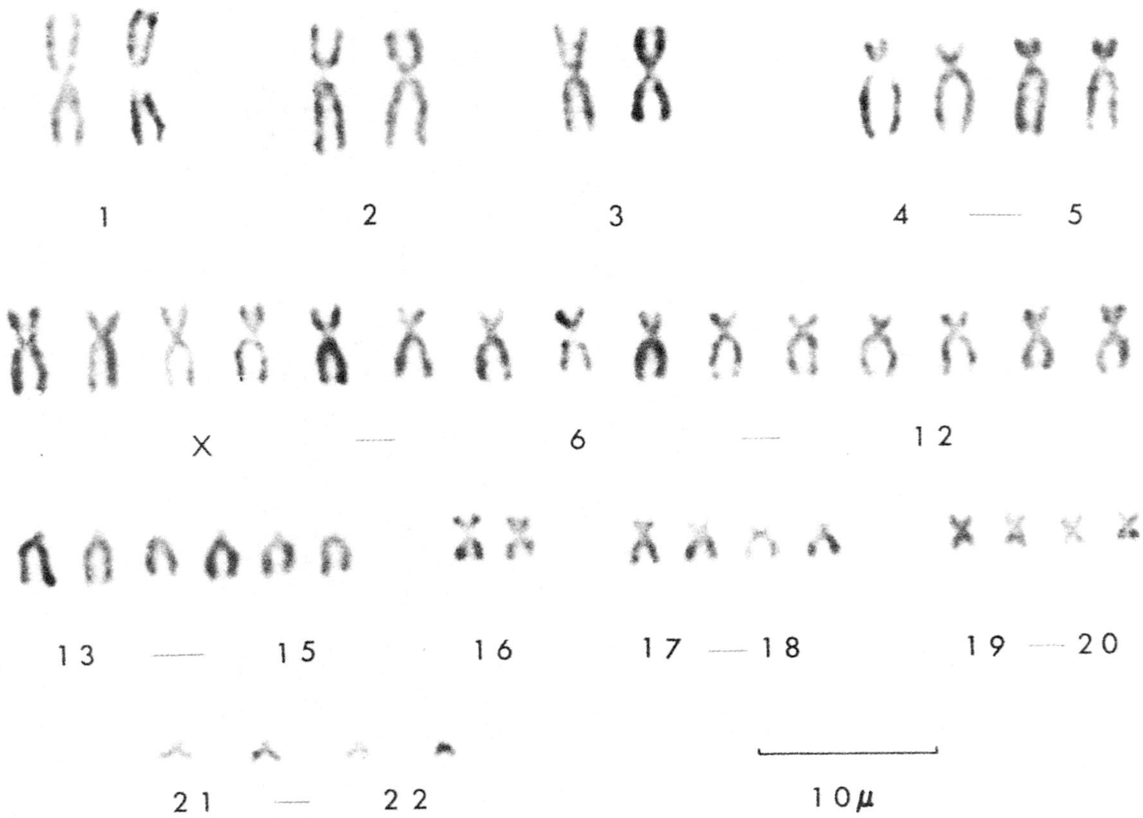

Fig. 4.9a. Karyotype from Fig. 4.9b.

4 Monosomy X (Figs. 4.9a and 4.9b).

This abnormality is associated with the well-defined clinical condition of Turner's syndrome. The most frequent features are primary amenorrhoea, short stature, rudimentary gonads and sterility. Other characteristics may include mental retardation, webbing of the neck, cubitus valgus and coarctation of the aorta.

Where all the cells in an individual are of XO constitution, then the abnormality may have arisen as the result of non-disjunction of types 1, 2 or 4 (p. 17 *et seq*) or by anaphase lag either at meiosis or early in the mitoses of embryonic development. Where it occurs in only some cells of the body the constitution of the other cell line may indicate which form of upset produced the monosomy. For example, if the mosaic is of XO and XX cells only, it probably arose as the result of anaphase lag. If the mosaic is XO/XX/XXX, it is likely to have arisen as a result of non-disjunction.

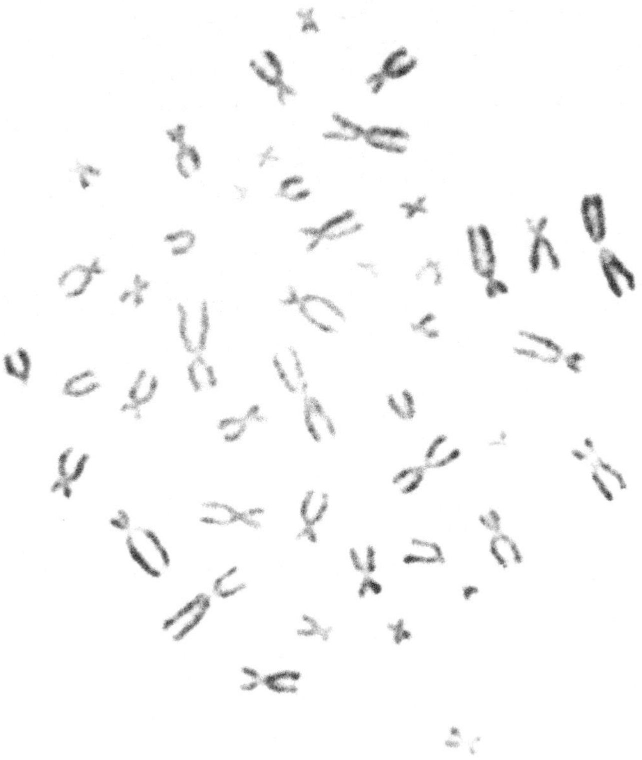

Fig. 4.9b. Metaphase figure showing monosomy X (45 chromosomes).

5 XXY (Figs. 4.10a and 410b).

This, the most common sex chromosome aberration, has been found in infertile males with small testes. The condition is called Klinefelter's syndrome. The extra X chromosome may be found in some or all of the cells of an individual.

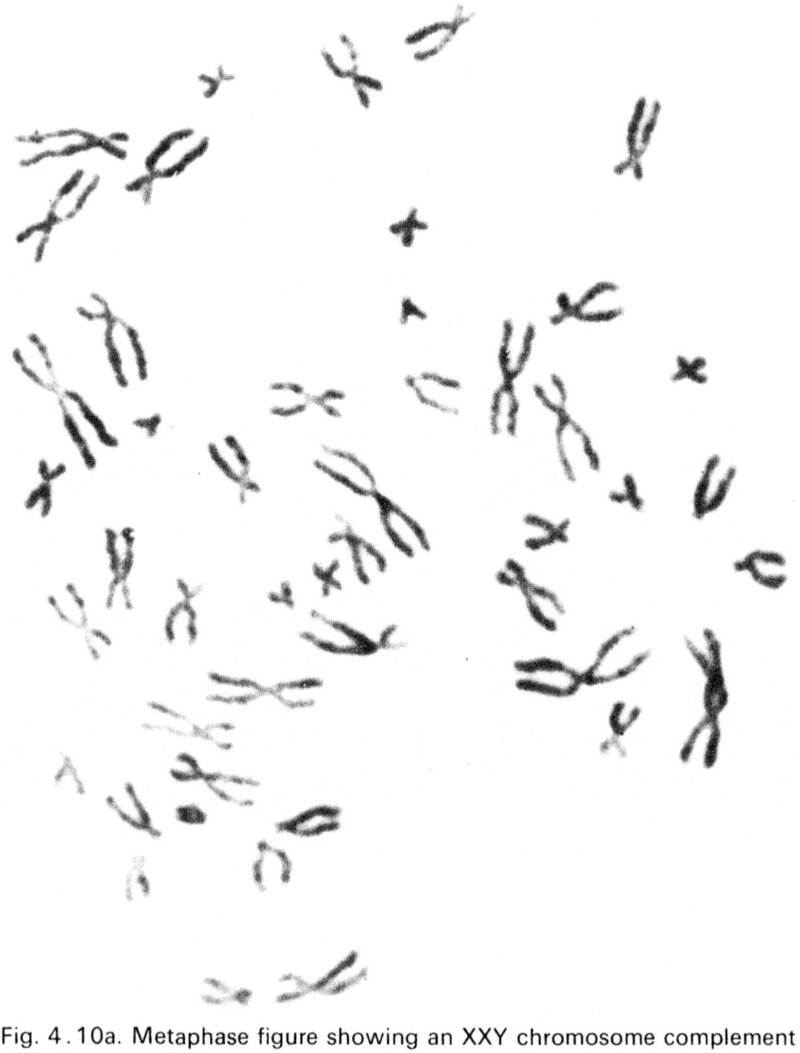

Fig. 4.10a. Metaphase figure showing an XXY chromosome complement (47 chromosomes).

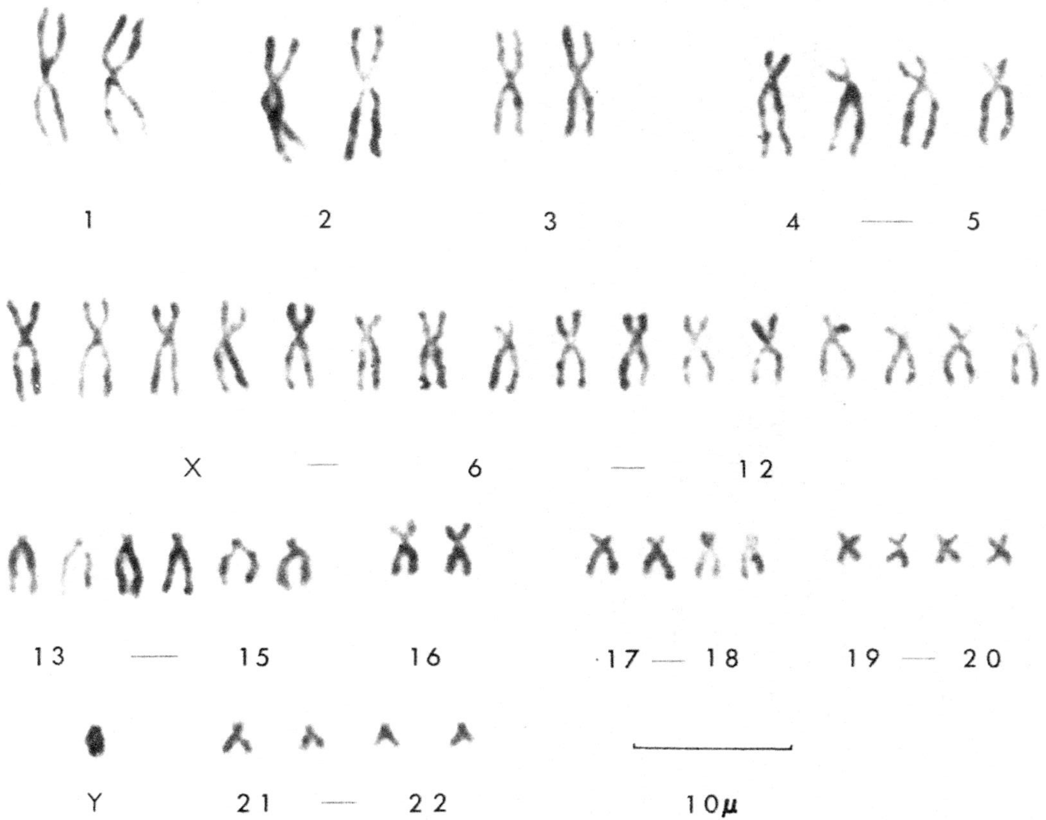

Fig. 4.10b. Karyotype from Fig. 4.10a.

6 XXXY (Fig. 4.11b).

When an individual has one or more extra members of any pair of chromosomes it is said to be polysomic for that chromosome. Polysomy for the sex chromosomes has frequently been observed in man and one such example is the XXXY chromosome complement. This karyotype was found in an individual, showing some features of Klinefelter's syndrome, with two cell lines, one with an XXXXY and one with an XXXY sex chromosome complement. It is therefore probable that three successive non-misjunctions of an XY zygote have contributed to this abnormality (Fig. 4.11a).

Another form of polysomy for the sex chromosomes is XYY. This has recently been observed with high frequency in groups of institutionalized males and appears to be associated with aggressive and antisocial behaviour. Many of these individuals are exceptionally tall.

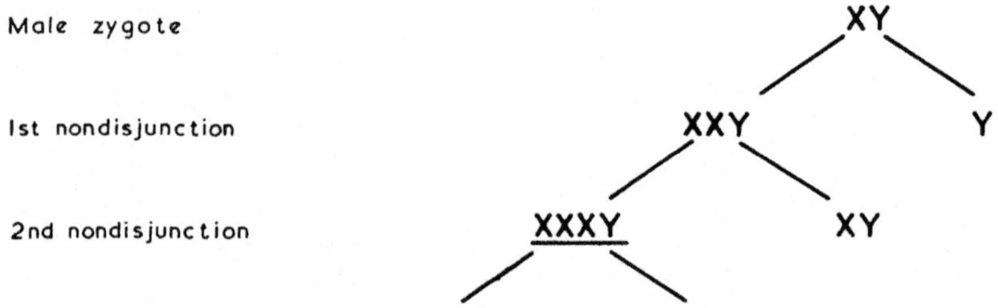

Fig. 4.11a. Possible derivation of an XXXY cell line. The cell lines underlined in the diagram are those which were observed in this mosaic.

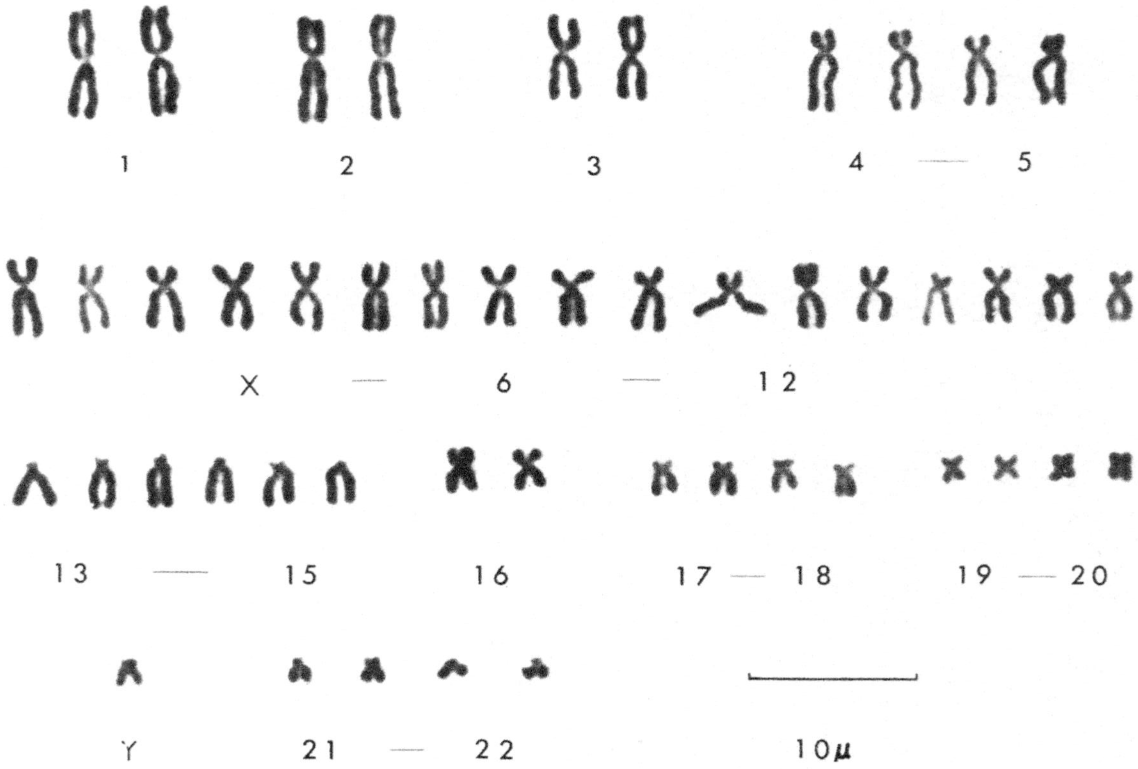

Fig. 4.11b. Karyotype demonstrating XXXY cell line with 48 chromosomes and 17 in the X–6–12 group.

7 XXYY (Fig. 4.12b).

Both sex chromosomes may disjoin simultaneously or successively to give duplication of the sex chromosome complement. This karyotype was observed in a hermaphrodite who was shown to be a mosaic of XXYY and XX cells (Fig. 4.12a). As well as the illustrated example of an XXYY chromosome complement examples of an XXXX type have been reported.

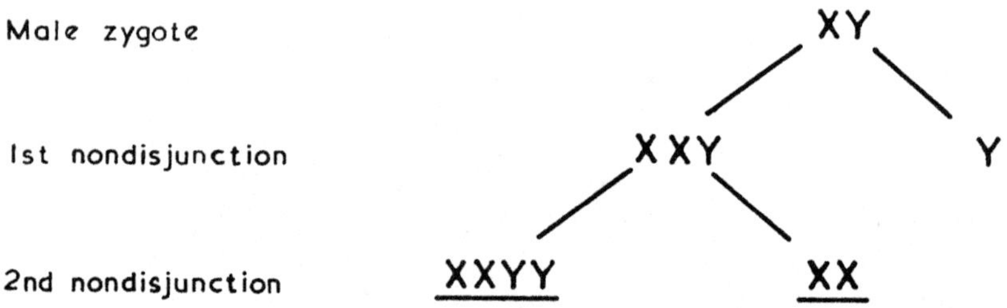

Fig. 4.12a. Possible derivation of XXYY cell line. The cell lines underlined in the diagram are those which were observed in this mosaic.

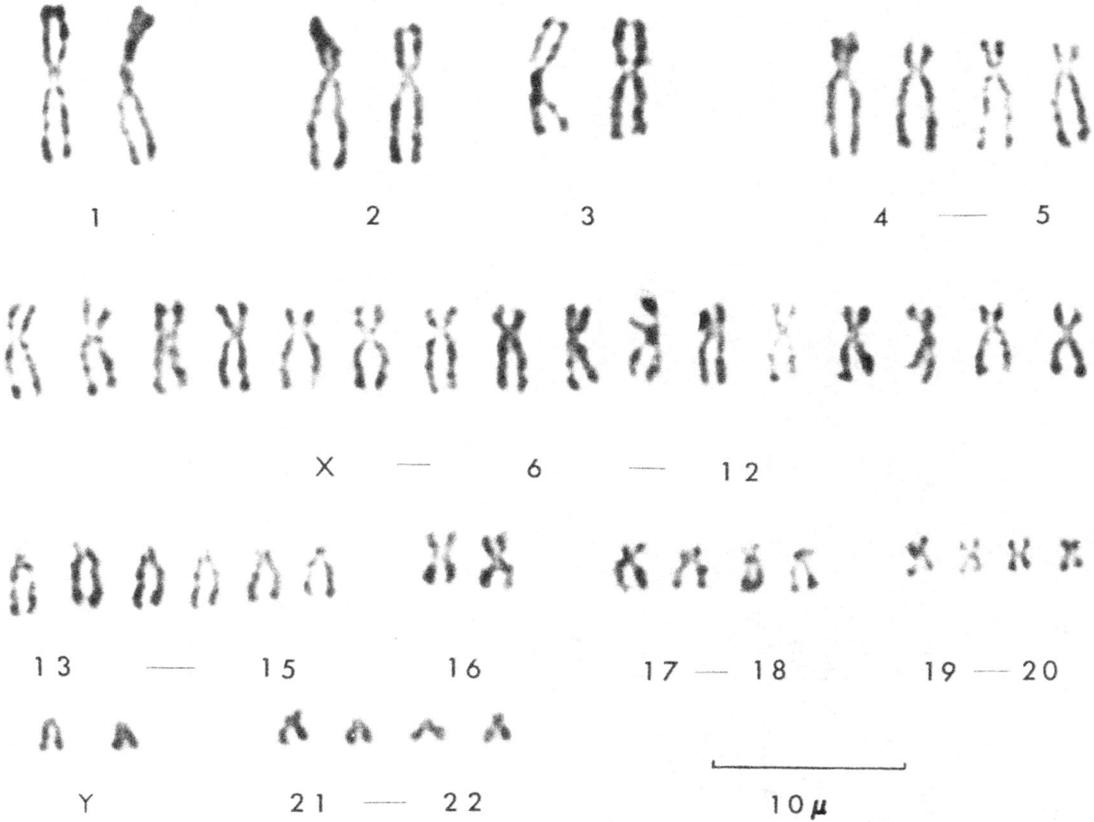

Fig. 4.12b. Karyotype demonstrating XXYY cell line with 48 chromosomes, 16 in the X–6–12 group and 2 Y chromosomes.

Chapter 5

Structural change

Upsets in the chromosome complement may be produced by structural change. Compared with numerical aberrations, detectable structural changes are relatively rare. The main forms of change encountered in humans are translocation, deletion, duplication and ring formation. The most important is translocation.

A translocation is the result of exchange of chromatin between chromosomes following breakage of these chromosomes. The two products of such a change (which are called T1 and T2) may look different from the normal homologues, but if the amount of chromatic material exchanged is equal in size, the translocation cannot be detected morphologically (Fig. 5.1). Where both products of translocation are present in an individual no deleterious effects ensue since all the genetic information required by the cell is present even if it is in the wrong position. The translocation is present in its balanced form and the individual is known as a translocation carrier.

A translocation carrier can produce four types of gamete as a result of chromosome reassortment during meiosis. From Fig. 5.2 it can be seen that, on fertilization by a normal gamete, only one of these four gametes is capable of producing a normal zygote. One will produce a zygote with a chromosome complement similar to that of the original carrier. In the two cases where the translocation product of one pair and the normal chromosome from the other pair pass to the daughter cell, zygotes are formed with the translocation in its unbalanced form. These zygotes carry extra material from one of the affected chromosomes and lack of material from the other. Such zygotes may develop into individuals with associated mental and physical maldevelopment.

Selection may operate against the abnormal gametes or against the abnormal embryos, but even so the chance of a translocation carrier producing an abnormal child is much higher than in the general population. The actual chance in

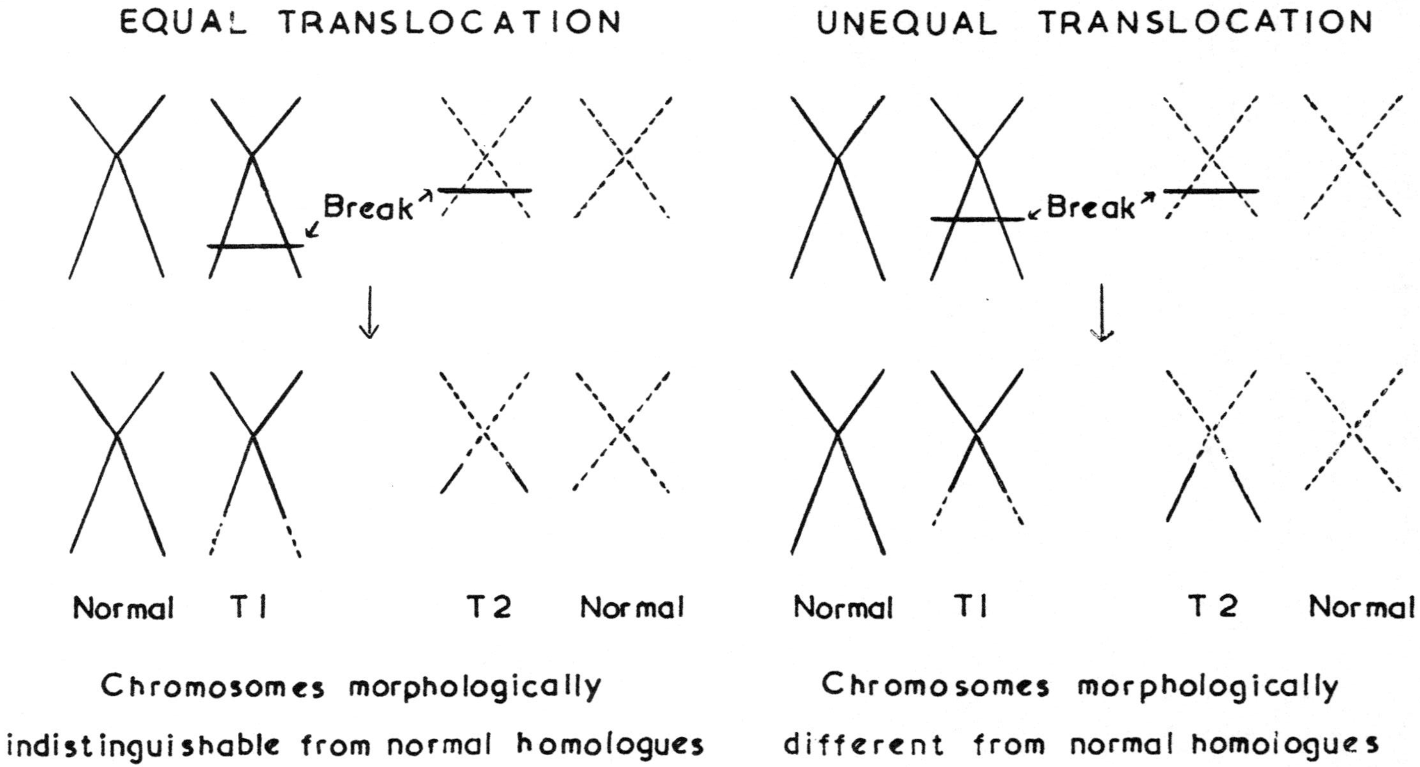

Fig. 5.1. The formation of translocation chromosomes as a result of exchange of chromatin segments of equal and unequal size.

any specific case can only be estimated from the observed frequency of affected births in families with a known translocation of the same type.

A translocation frequently found in human populations is one involving an exchange of material between a chromosome of the 13–15 group and a chromosome of the 21–22 group. Conventionally the chromosomes involved are referred to as a Number 15 and a Number 21 (Fig. 5.3a). Of the four

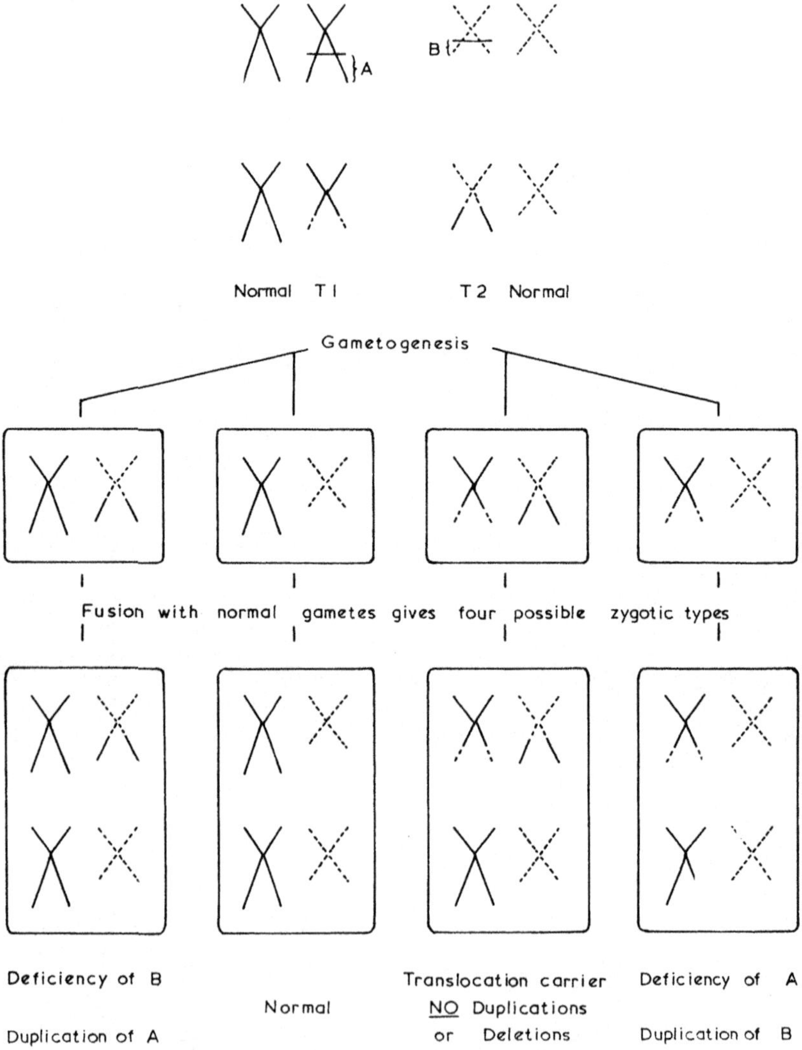

Fig. 5.2. The formation and transmission of an unequal translocation from parent to offspring.

possible products only one is normal. One is deficient in 21 material and has been presumed to be inviable since no examples of this type of complement have been recovered. One has a karyotype similar to that of the original carrier (Fig. 5.3b). The fourth has an excess of 21 material amounting to almost complete trisomy for a member of the 21 pair (Fig. 5.3c). This condition is associated with mongolism.

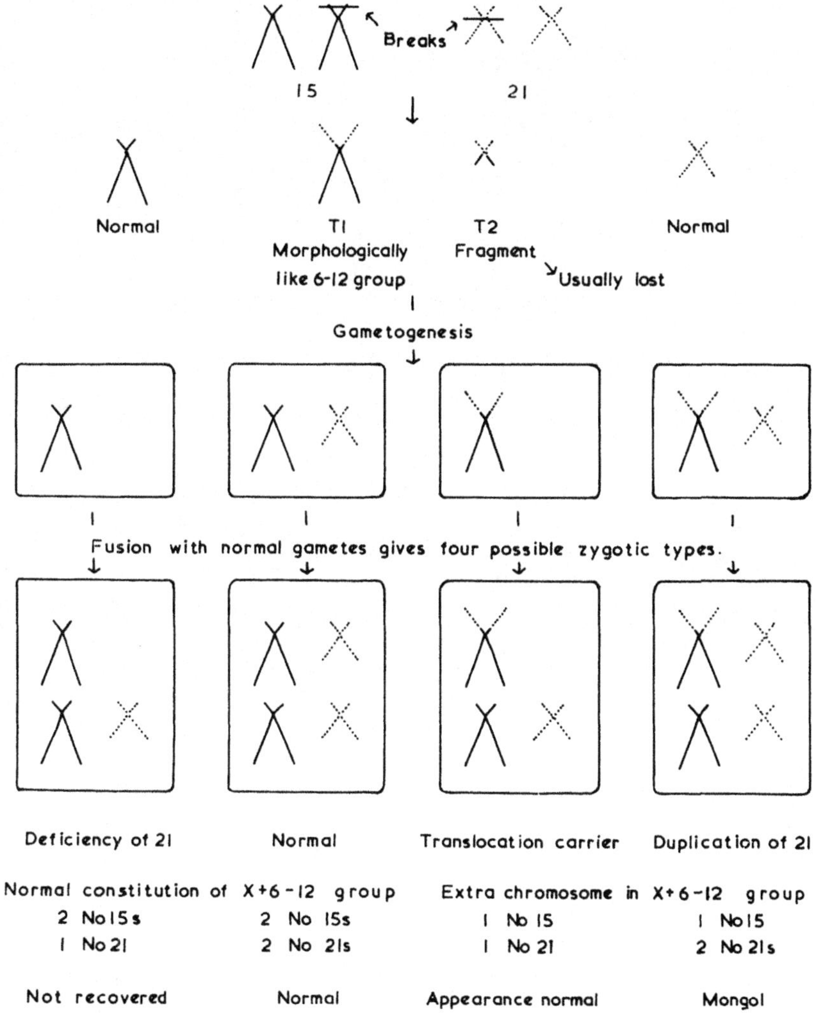

Fig. 5.3a. The formation and transmission of a translocation between a Number 15 and a Number 21 chromosome.

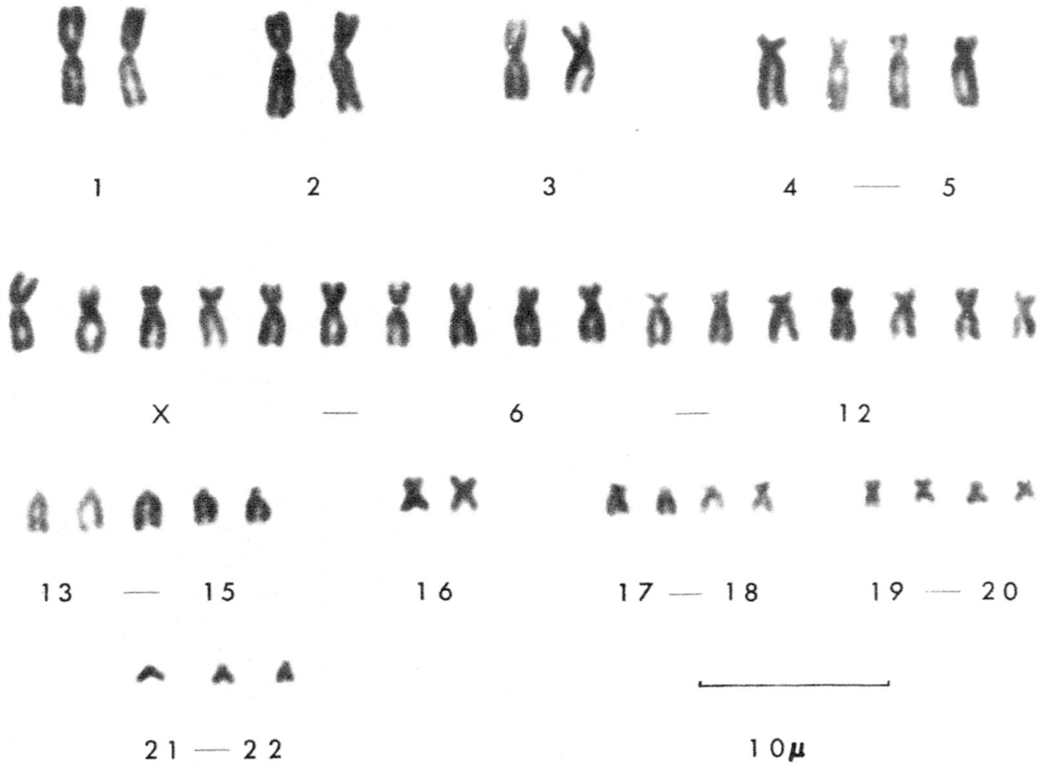

Fig. 5.3b. Karyotype of a carrier of a 15/21 translocation. There are only 5 in the 13–15 group, 3 in the 21–22 group and an extra member of the X–6–12 group (45 chromosomes).

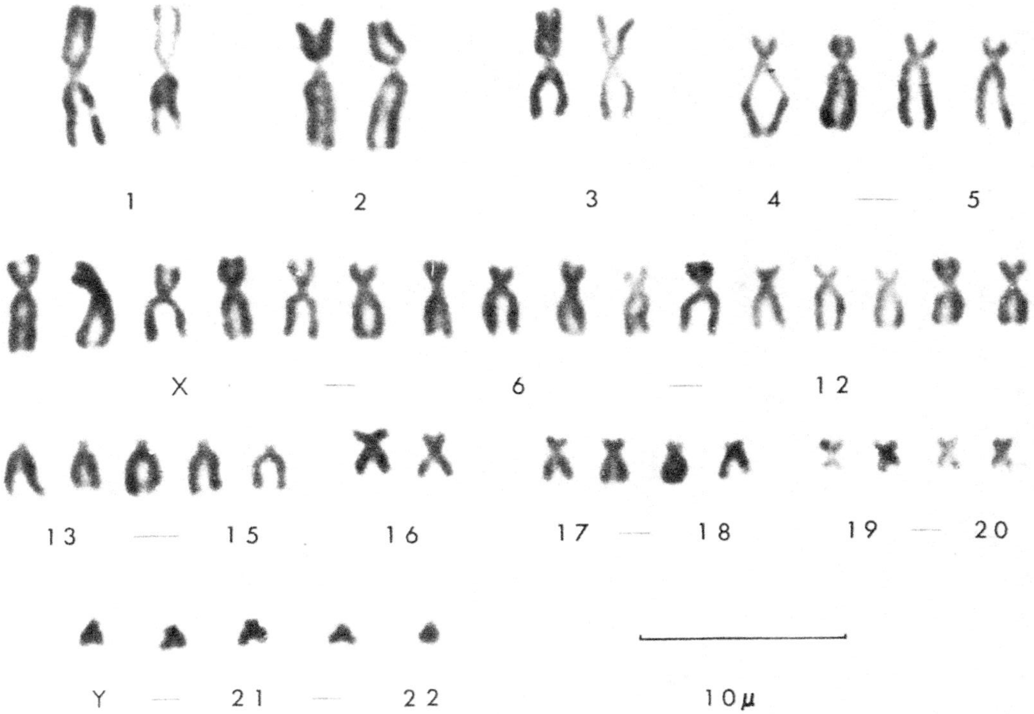

Fig. 5.3c. Karyotype of an affected individual in which the 15–21 translocation is present in its unbalanced form. There are only 5 in 13–15 group, 4 in the 21–22 group and an extra member of the X–6–12 group (46 chromosomes).

A translocation may be carried in its balanced form (Fig. 5.4a) for several generations before an abnormal individual is produced with the translocation in its unbalanced form. A translocation between a Number 2 chromosome and a member of the 6–12 group has been observed in three generations of a family. Fig. 5.4b shows the derivation of this translocation and Fig. 5.4c is a family tree illustrating the way in which the translocation has been transmitted.

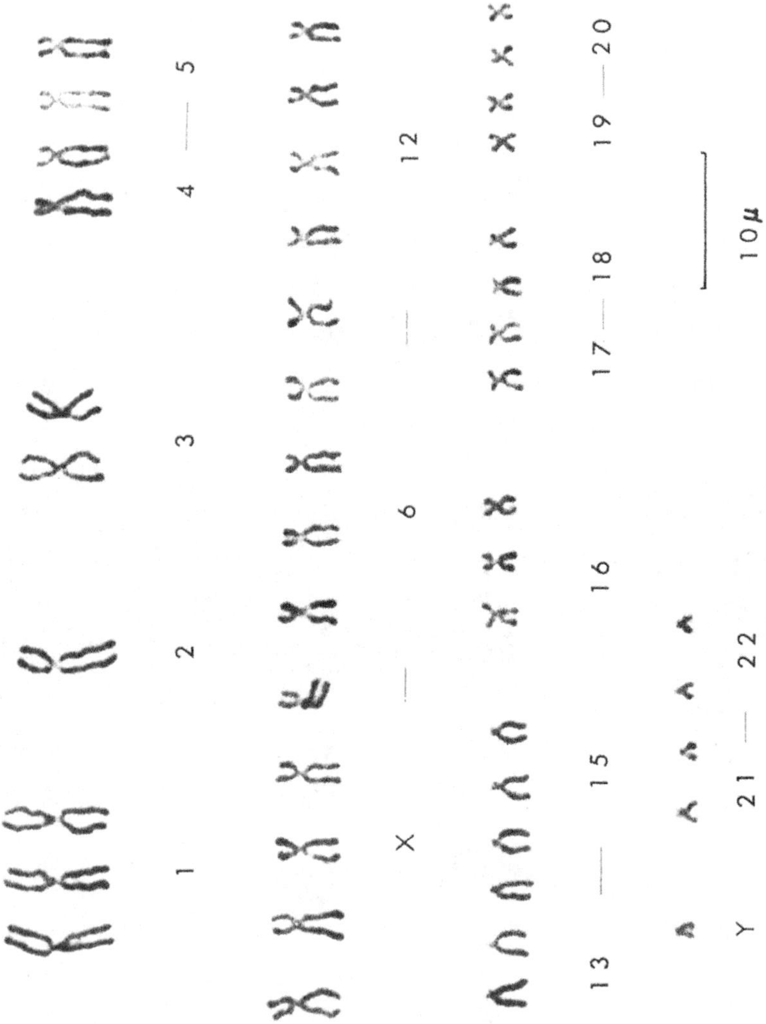

Fig. 5.4a. Karyotype of a carrier of a 2/6–12 translocation.

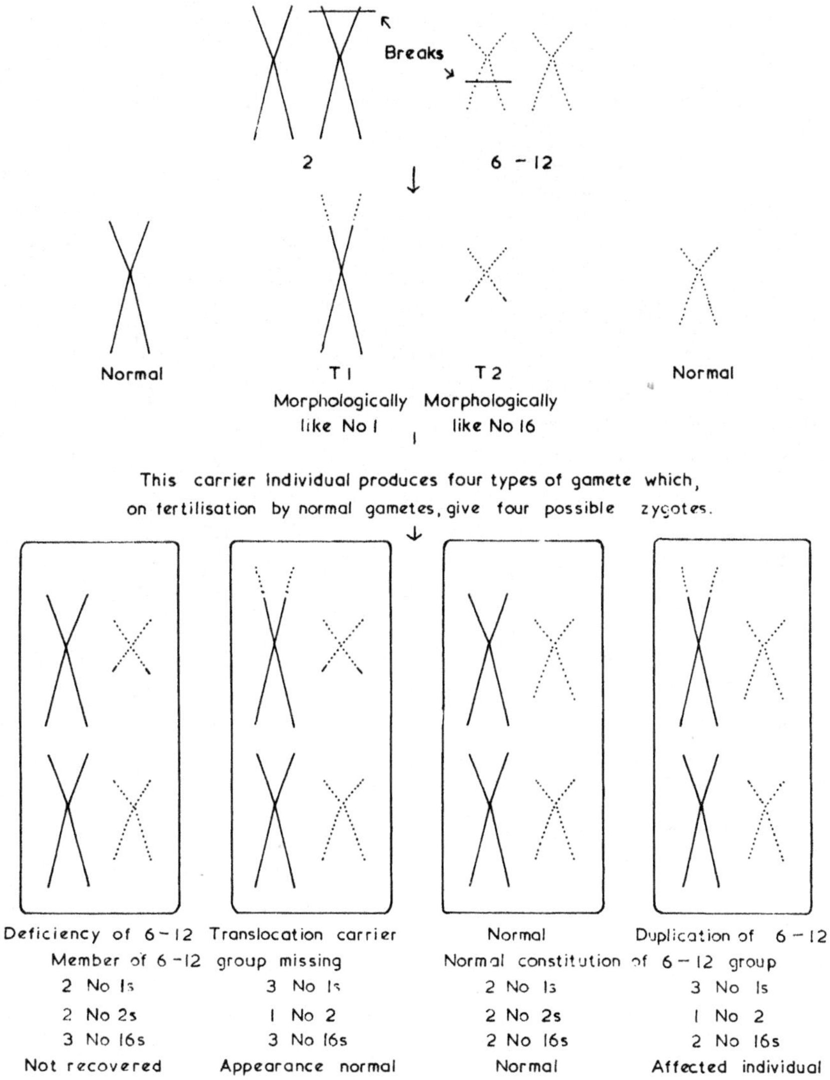

Fig. 5.4b. The formation and transmission of a translocation between a Number 2 chromosome and a member of the X–6–12 group.

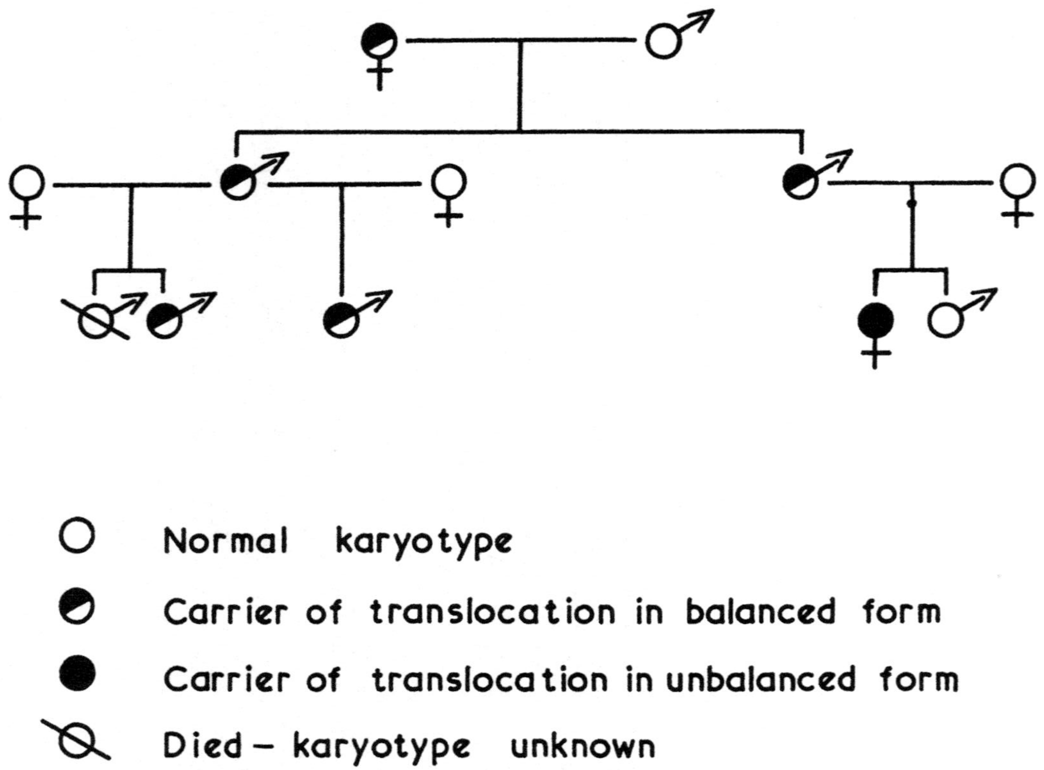

Fig. 5.4c. A family tree showing how the 2/6–12 translocation has been transmitted through three generations of a family.

Deletions are a common form of structural alteration which have been observed in members of several groups, e.g. 4–5, 13–15, 17–18 and 21–22. A small deletion is very difficult to detect on length alone due to the presence of variation in the overall length of chromosomes. It is only when the deletion alters the relative position of the centromere in the affected chromosome that the change becomes apparent. Where a deletion is suspected care must be taken to exclude the possibility of unequal translocation between the affected chromosome and an unknown partner. One of the most commonly found deletions is partial or almost complete loss of the short arms of a Number 5 chromosome (Fig. 5.5). Associated with this anomaly is the syndrome known as 'Cri du chat'. A characteristic of this syndrome is that the cry of affected individuals is high pitched like the mewing of a cat. Non-specific abnormalities also observed include mental subnormality, micrognathia and low-set ears.

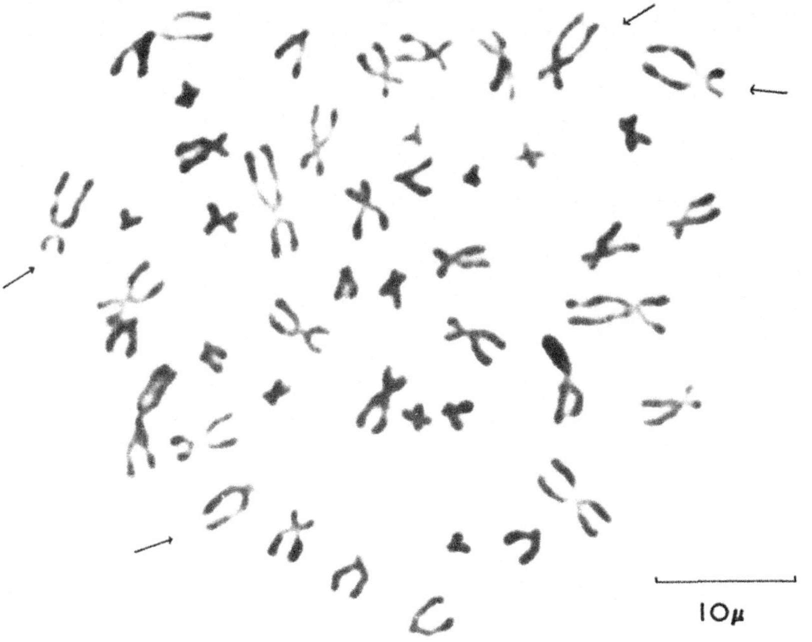

Fig. 5.5. Metaphase figure showing a deletion of the short arms of a Number 5 chromosome. The members of the 4–5 group are arrowed.

The form of chromosomal duplication most commonly encountered in man is that produced by isochromosome formation. Isochromosomes are produced when, during cell division, a centromere undergoes an abnormal form of division. Two isochromosomes are produced as the result of such misdivision, one with complete duplication of the short arm of the chromosome and one with duplication of long arm material. Often cells with one type of isochromosome are more viable than those with the other. Isochromosomes for the long arm of the X chromosome have been observed more frequently than isochromosomes for the short arms. A characteristic of isochromosomes is their precise metacentricity, and they may often be distinguished from translocation products by this criterion. An isochromosome for the long arms of the X is morphologically indistinguishable from the Number 3 chromosomes (Fig. 5.6) but may be identified by its characteristic behaviour when studied autoradiographically (Chapter 6). Individuals with an isochromosome of this type are usually mosaics with an XO cell and show similar features to XO individuals.

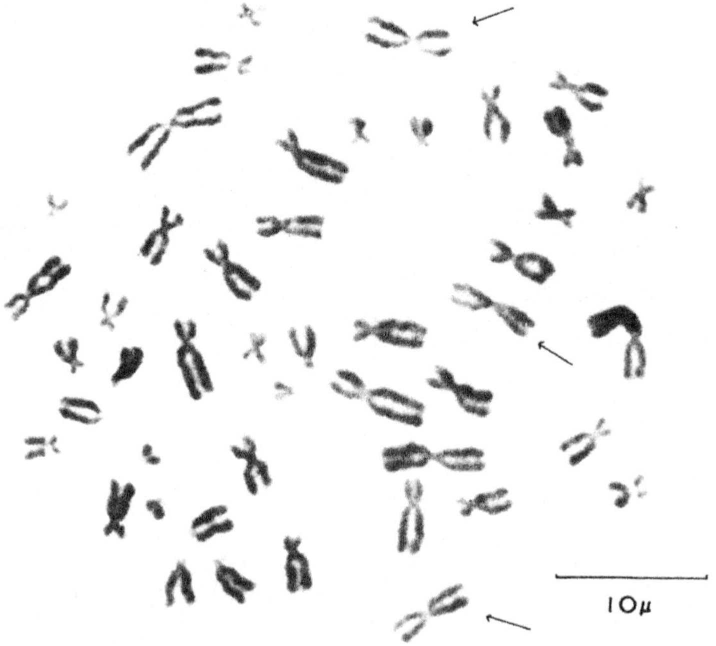

Fig. 5.6. Metaphase figure showing apparently three Number 3 chromosomes, one of which has been shown by autoradiography to be an isochromosome X.

A relatively unusual form of chromosomal change is ring chromosome formation. Ring chromosomes may be formed when terminal deletions within the same chromosome are followed by a fusion of the exposed ends. Within an individual the ring may be variable in size and genetic constitution as a result of abnormal division. Therefore superimposed on the original deletion which gave rise to the ring, daughter cells may show deletion and/or duplication for other parts of the chromatic material. The smaller the ring is, the more likely it is to be lost during division. When the ring chromosome is derived from a relatively small chromosome, such as the X chromosome or a member of the 21–22 group, the affected individual is likely to be a mosaic of cells, some with the ring chromosome, some without. A ring derived from a Number 1 chromosome has been observed in a case of microcephalic dwarfism (Fig. 5.7). This, however, was a comparatively large ring and was present in one form or another in all the cells of the individual.

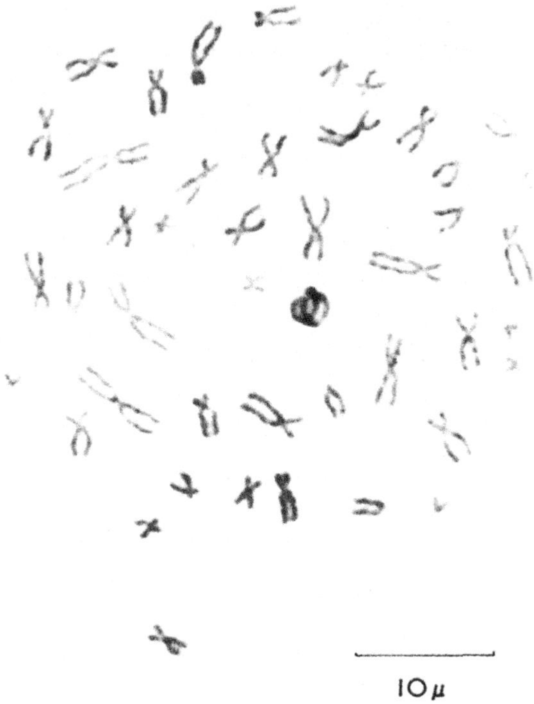

Fig. 5.7. Metaphase figure showing a ring derived from a Number 1 chromosome.

Chapter 6

The X chromosome

One of the X chromosomes in the normal female behaves differently from the other chromosomes. It becomes very condensed during interphase, forming a darkly-staining mass called the sex chromatin body. This body is about 1 μ in diameter and often lies against the nuclear membrane. There is no sex chromatin body in the normal male and so this property of the X chromosome may be used to indicate the nuclear sex in an individual.

Sex chromatin may be studied in a variety of tissues. The tissues most generally used are the surface epithelial cells of the buccal mucosa and fibroblasts grown in skin culture. A buccal smear is scraped from the inside of the mouth and smeared onto a slide. The cells are then fixed, hydrolysed in acid to remove bacteria, and may be stained in cresyl fast violet acetate.

The number of sex chromatin bodies present is one less than the number of X chromosomes. In an XO individual (p. 29) there is no sex chromatin body. In a XXX or XXXY individual (p. 27, 32) there are two bodies (Fig. 6.1) and in a XXXX or XXXXY individual there are three bodies (Fig. 6.2). Abnormal X chromosomes of large size (p. 46) are found to give rise to large bodies and deleted X's produce small bodies.

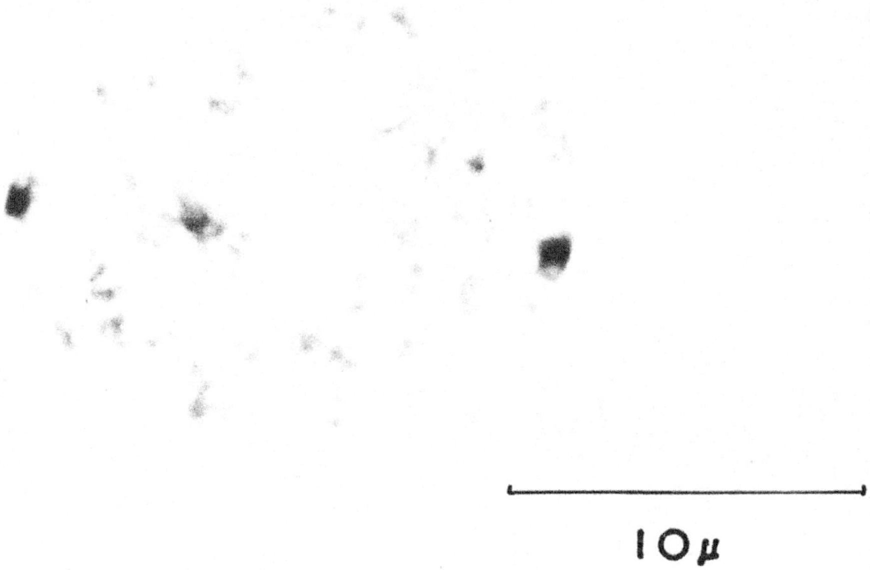

Fig. 6.1. Nucleus with two sex chromatin bodies.

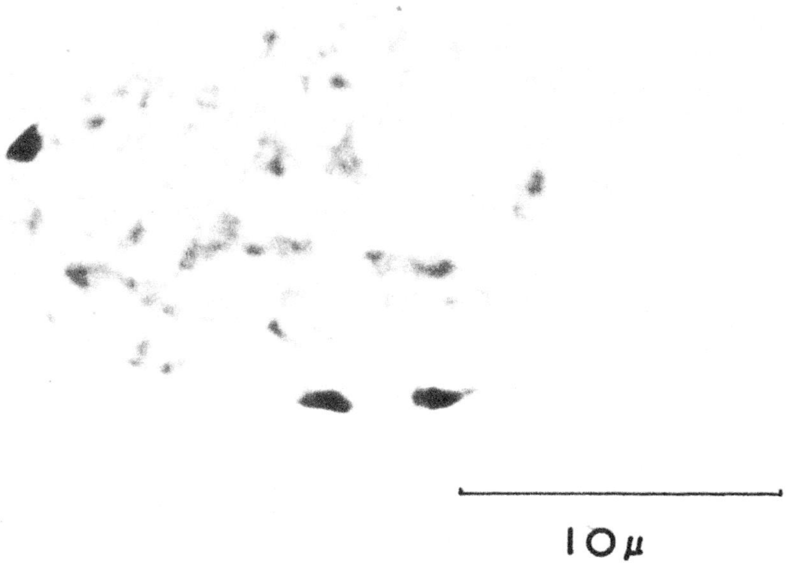

Fig. 6.2. Nucleus with three sex chromatin bodies.

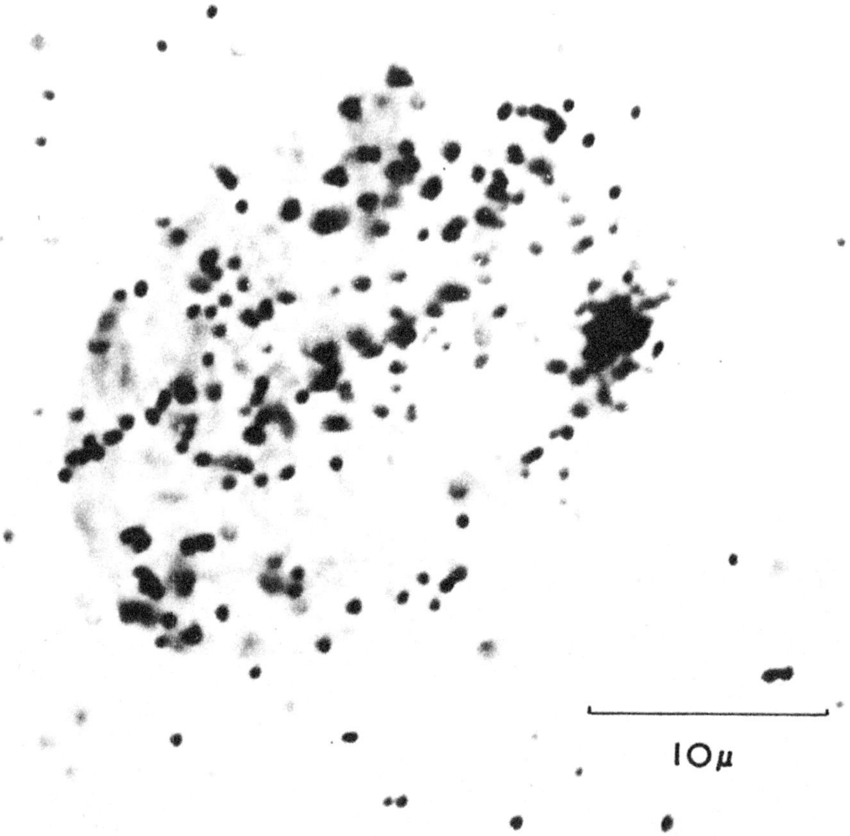

Fig. 6.3a. Nucleus showing heavy pattern of radioactive labelling over the sex chromatin body.

Another way in which the X chromosome behaves differently is in its replicating pattern as shown by labelling with tritiated thymidine. In the normal female one of the X chromosomes replicates later than the rest of the chromosome complement. Therefore in cells supplied with the isotope at

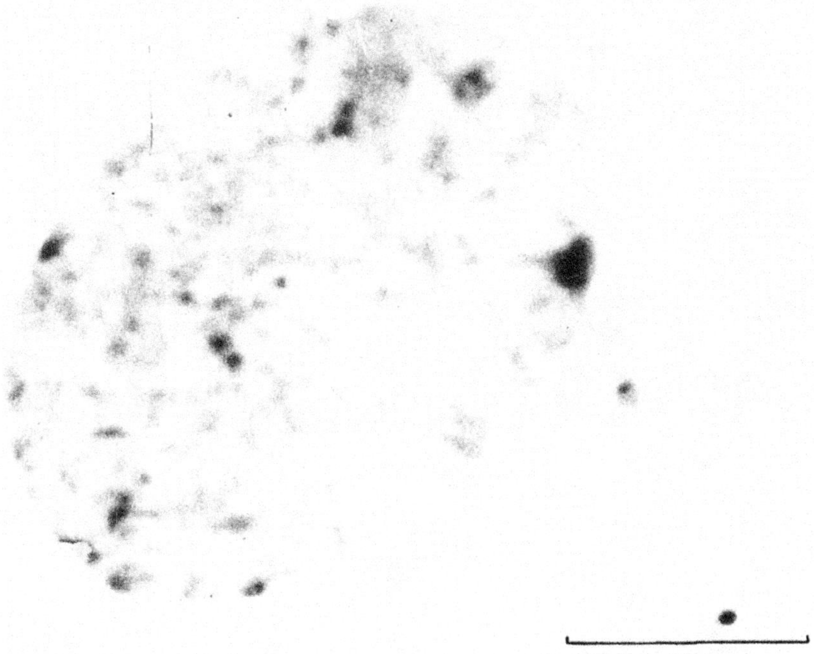

Fig. 6.3b. The same nucleus after removal of the autoradiographic film.

the end of the period of DNA synthesis, the X is apparent as a heavily labelled sex chromatin body at interphase (Figs. 6.3a and 6.3b) and a heavily labelled chromosome at metaphase (Figs. 6.4a and 6.4b). This method may be used to identify one of the X chromosomes. In instances where a structurally abnormal chromosome is present, in the X–6–12 group, it may be identified as an X by this autoradiographic technique.

Fig. 6.4a. This cell was supplied with tritiated thymidine for four hours before harvesting and the X is apparent as the most heavily labelled chromosome.

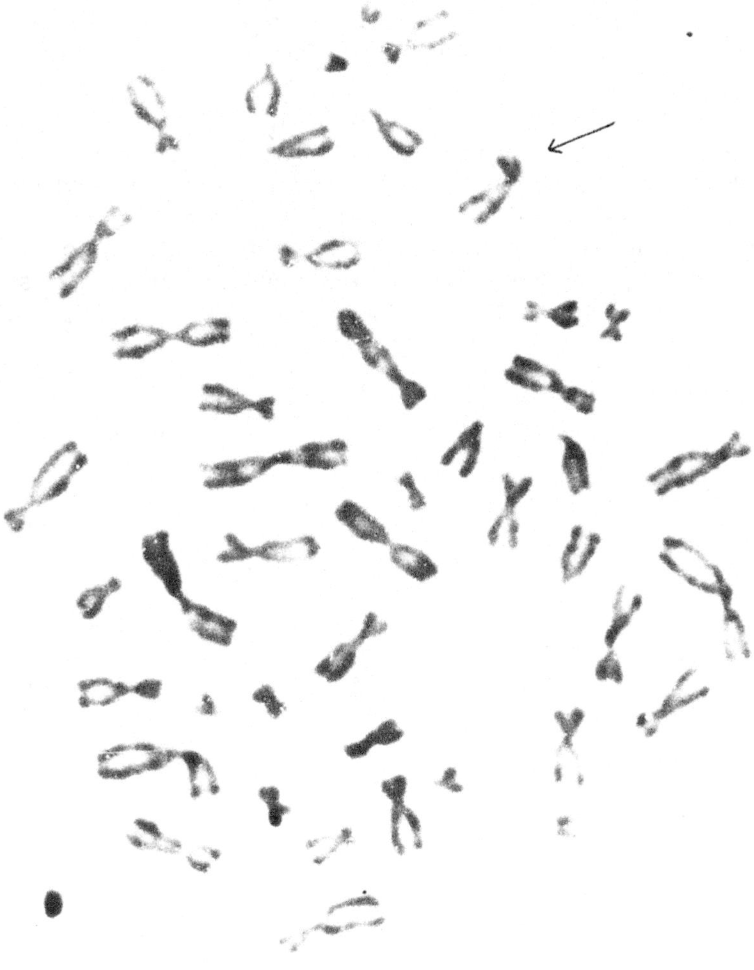

Fig. 6.4b. The same cell after removal of the autoradiographic film.

The third interesting feature of the X chromosome is its genetic inactivity. It has now been shown that in the somatic cells of the female part of one of the X chromosomes is genetically inactive. This inactivation occurs early in embryonic life and it is at random so that, either the maternal or paternal X may be affected in a particular cell. This explains the phenomenon of dosage compensation observed in man where the products of certain X linked genes are quantitatively the same in the homozygous female with two Xs and the hemizygous male with one X.

It has been observed that monosomy or trisomy for an autosome is either lethal or produces severe abnormalities. However monosomy, trisomy and even higher polysomy of the X are tolerated by the individual. It is probable that, because part of one of the X chromosomes in the normal female is genetically inactive, there is relatively little effect on an individual if a whole X chromosome is missing or present in excess.

Glossary

Primary Amenorrhoea.	Absence of menstruation.
Brushfield Spots.	Non-pigmented spots on the iris.
Coarctation.	Narrowing.
Cubitus valgus	Increased carrying angle of the forearm.
Dermatoglyphics.	The study of the ridge patterns on the soles of the feet and the palms of the hands.
Epicanthic folds.	Vertical folds of skin where the top and bottom eyelid meet at the inner edge.
Hermaphrodite (true).	An individual with both ovarian and testicular tissue.
Hypotonia.	Diminished muscle tone.
Micrognathia.	Underdevelopment of the mandible.
Occiput.	The back of the head.
Palpebral fissures.	The opening between the eyelids.

Index

Acrocentric 11
Amenorrhoea 29
Anaphase 4
Anaphase lag 21
Autoradiograph 16
Autosomes 11
Bivalent 4
Blood 8
Bone Marrow 8
Brushfield spots 23
Buccal mucosa 48
Cell 1
Centromere 4
Chiasma 4
Chick embryo extract 9
Chromatids 2
Chromatin 12
Chromosomes 2
Colcemid 8
Cresyl fast violet acetate 48
Cri du chat 45
Crossing over 4
Cubitus valgus 29
Cytoplasm 1
Deletion 45
Deoxyribonucleic acid (DNA) 1, 2
Dermatoglyphics 23
Diploid 2
Epicanthic folds 23
Gamete 2
Genes 1
Haploid 2
Heparin 8
Hermaphrodite 34
Homologues 2
Hypotonia 23
Hypotonic 8
Interphase 1
Isochromosome 46
Karyotype 11
Klinefelter's syndrome 30
Lacto-aceto-orcein 1, 8
Leukaemia 9
Medium 8
Meiosis 4
Metacentric 10
Metaphase 4
Micrognathia 24
Mitosis 2
Mongolism 23
Monosomy 17
Mosaicism 21
Non-disjunction 17
Nuclear membrane 1
Nuclear sap 1
Nucleus 1
Occiput 23
Palpebral fissures 23
Phytohaemaglutinin 8
Poles 4
Polysomy 32
Prophase 2
Protein synthesis 1
Red cells 8
Replication 2
Ribonucleic acid (RNA) 1
Ring chromosomes 47
Satellites 15
Secondary constrictions 14
Serum 9
Sex chromatin body 48
Sex chromosomes 11
Skin 8
Spindle 4
Submetacentric 10
Telophase 4
Translocation 36
Translocation carrier 36
Trisomy 17
Tritiated thymidine 16
Trypsin 9
Turner's syndrome 29
White cells 8
Zygote 2

See Glossary

Printed in Dunstable, United Kingdom

91221443R00037